Electronic Materials: Science & Technology

Series Editor

Harry L. Tuller, Cambridge, MA, USA

The series *Electronic Materials: Science and Technology* will address the following goals:

- Bridge the gap between theory and application
- Foster and facilitate communication among materials scientists, electrical engineers, physicists and chemists
- Provide publication with an interdisciplinary approach in the following topic areas:
- Sensors and Actuators
- Electrically Active Ceramics and Polymers
- Structure-Property-Processing-Performance Correlations in Electronic Materials
- Electronically Active Interfaces
- High Tc Superconducting Materials
- Optoelectronic Materials
- Composite Materials
- Defect Engineering
- Solid State Ionics
- Electronic Materials in Energy Conversion
- Solar Cells, High Energy Density Microbatteries, Solid State Fuel Cells, etc.

Masahiro Hiramoto

Organic Photocurrent Multiplication

Springer

Masahiro Hiramoto
Department of Materials Molecular Science
Institute for Molecular Science
Okazaki, Aichi, Japan

ISSN 1386-3290
Electronic Materials: Science & Technology
ISBN 978-981-99-1236-0 ISBN 978-981-99-1237-7 (eBook)
https://doi.org/10.1007/978-981-99-1237-7

This Springer imprint is published by the registered company Springer Nature Singapore Pte Ltd.
The registered company address is: 152 Beach Road, #21-01/04 Gateway East, Singapore 189721, Singapore

Preface

God is in the detail. Actually, in the nano-world at the organic/metal interface, undoubtedly, God is in the detail. This book describes the adventurous life of one scientist who was fascinated by the organic/metal interface.

In 1992, I encountered a novel phenomenon of "photocurrent multiplication in organic semiconductors" and continued this line of research until 2008, a total of more than 16 years. Over this time period, photocurrent multiplication was gradually unveiled to be caused by the concrete nanostructures of molecular steps and kinks on organic crystal surfaces. Macroscopic device functions and molecular nanostructure are directly linked to this phenomenon.

I believe that original ideas are the most important aspect of scientific research. This can be a lonely journey that includes the conceiving of new ideas, logical thinking, and reflection of my way of life. The life of the author offers useful references for the next generation of scientists. Thus, I decided to write this book.

In Chaps. 1–5, I describe the discovery of multiplication and solving the riddle of its origin. In Chaps. 6–9, I describe applications of multiplication devices. In Chap. 11, I survey the progress of multiplication devices after finishing my studies, and in Chaps. 10 and 12, I propose future advances in multiplication devices.

I switched my field from inorganic to organic semiconductors in 1988 at the age of 30. I invented the blended junction (bulk heterojunction) in 1991 and the tandem junction in 1990 for organic solar cells. Simultaneously, I discovered the organic photocurrent multiplication described herein in 1992. My most important studies were performed between the ages of 30 to 34. Figure 1 is a photograph from the Gordon Research Conference, Andover, 1994. I remember nostalgically the first discussion with Prof. C. W. Tang regarding his famous work on organic solar cells. The researchers of that generation knew my work well and many discussions occurred over a few beers. The first half of my 30s were my most productive and are critically important for young researchers. Creativity is strongly stimulated after switching fields. I expect that this book, which describes a lonely research journey, will serve as a good guide for young researchers.

Fig. 1 Photograph from the Gordon Research Conference on "Electronic Processes in Organic Materials", Andover, 1994. A mixture of top international scientists and young researchers

I appreciate very much the 18 Osaka University students, namely, T. Miyao, K. Yoshimura, T. Imahigashi, S. Kawase, T. Katsume, K. Nakayama, I. Sato, H. Kumaoka, Y. Niguma, M. Yoshida, A. Miki, G. Matsunobu, Y. Oishi, K. Fujino, M. Naito, K. Yamauchi, Y. Motohashi, and K. Kitada, who performed various experiments in support of this research.

My scientific carrier was strongly motivated by my deceased mother, Shizuko (calm princess), and father, Masanori (law of correctness). It was also strongly motivated by my daughters, Megumi (blessed and gracious) and Mika (beauty and wisdom), and my son, You-dai (open and magnificent). Here, in the parentheses, the English meaning of the Japanese names is indicated. Figure 2 shows a photograph of the Hiramoto family in our homestead house in Hiroshima.

I dedicate this book to my parents and my children.

I proposed the outline of this book in the summer of 2020 to Shinichi Koizumi, Ph.D., Senior Editor, Books Chemistry & Materials Science. It took more than one year to complete this book. I appreciate Shinichi Koizumi for his encouragement and assistance and I also appreciate my secretary, Ms. Yasuko Okura, for her assistance with obtaining copyright permissions.

Fig. 2 Photograph of the Hiramoto family in their homestead house in Hiroshima. The deceased Shizuko and Masanori are in the portraits (upper left)

I strongly hope that the young scientists of the next generation will develop this field vigorously.

平本昌宏

Okazaki, Aichi, Japan
December 2022

Masahiro Hiramoto

Contents

1 Photocurrent Multiplication in Amorphous Silicon Carbide Films 1
1.1 Photocurrent Multiplication in Amorphous Silicon Carbide Films 1
1.1.1 Background 1
1.1.2 Motivation 2
1.1.3 Discovery of Photocurrent Multiplication 2
1.1.4 Spectral Dependence 3
1.1.5 Light Intensity Dependence 4
1.1.6 Transient Response 4
1.1.7 Interfacial Structure 5
1.1.8 Multiplication Mechanism 7
1.1.9 Temperature Dependence 8
1.2 Photo-Modulation of Photocurrent Multiplication 8
1.2.1 Basic Idea 9
1.2.2 Multiplication Suppression 9
1.2.3 Suppression Wavelength Region 10
1.2.4 Suppression Mechanism 11
1.3 Conclusion 12
References 13

2 Photocurrent Multiplication in Organic Semiconductor Films 15
2.1 Discovery 15
2.1.1 An Original Idea: Taking Out the Interface 15
2.2 Photocurrent Multiplication in *n*-Type Organic Semiconductor Films 17
2.2.1 Multiplication Rate Reaching Ten Thousand-Fold 17
2.2.2 Identifying the Interface 19
2.2.3 Transient Response 19
2.2.4 Multiplication Mechanism—Tunneling Injection of Electrons 19

2.2.5 Hole Traps—Origin of Multiplication 21
2.3 Photocurrent Multiplication in *p*-Type Organic Semiconductor Films 21
2.3.1 Multiplication Behavior 22
2.3.2 Influence of the Kinds of Electrode Metals 22
2.3.3 Specifying the Interface 23
2.3.4 Transient Response 24
2.3.5 Multiplication Mechanism—Tunneling Injection of Holes 24
2.4 Generality 25
2.5 Room-Temperature Multiplication 28
2.5.1 Multiplication Rate 28
2.5.2 Spectral Sensitivity 28
2.5.3 Influence of the Kinds of Metals 29
2.5.4 Multiplication Mechanism 29
2.6 Conclusion 29
References 30

3 Analyses of Multiplication Behaviors—Structural Trap 33
3.1 Multiplication Mechanism 33
3.2 Transient Response 33
3.2.1 Two Components of Photocurrent 33
3.2.2 First Component: Primary Photocurrent 35
3.2.3 Second Component: Tunneling Injection 36
3.2.4 Direct Tracing of Multiplication Process 36
3.2.5 Required Charges for Multiplication 36
3.3 Fowler-Nordheim (FN) Analyses 37
3.3.1 FN Plots of Multiplied Photocurrent [6, 7] 37
3.3.2 Estimation of Field Concentration Width (d) 38
3.3.3 Estimation of Trap Density 40
3.4 TSC Measurements 40
3.4.1 Energetic Depth of Traps 40
3.4.2 Trap Depth Versus Electric Field 41
3.5 Structural Trap Model 43
3.5.1 Experimental Findings 43
3.5.2 Model Proposal 43
3.6 Clues for the Origin of Structural Traps 43
3.6.1 Deposition Rate 43
3.6.2 Material Dependence 44
3.6.3 Relationship Between Trap Density and Onset Charge 45
3.7 Conclusions 46
References 48

4 Morphology of Organic/Metal Interface and Photocurrent Multiplication Behaviors 49
4.1 Photocurrent Multiplication and Structural Trap Model 49
4.2 Morphology of Organic Films and Multiplication Behaviors 50
4.2.1 Motivation 50
4.2.2 Three Perylene Pigments with Different Side Chains 50
4.2.3 Multiplication Rates 51
4.2.4 Morphology of Organic Films 51
4.2.5 Crystallinity of Organic Films 51
4.2.6 Multiplication Induction by Crystallization 52
4.2.7 Energetic Barrier Height at Organic/Metal Junctions 53
4.2.8 Explanation Based on Structural Trap Model 55
4.2.9 Lateral Spatial Separation of Trap Sites and Injection Sites 55
4.2.10 Identification of Interfacial Crystallization Effect 56
4.3 Metal Morphology Observed by SEM and Multiplication Behaviors 58
4.3.1 Motivation 58
4.3.2 Effects of Au Deposition Conditions on Perylene Pigment Film [15] 58
4.3.3 Resistivity-Heated Deposition Versus Sputtering Deposition 59
4.3.4 Deposition Rate 59
4.3.5 Morphology of Au Films Observed by SEM 60
4.3.6 Explanation by Structural Trap Model 61
4.4 Metal Morphology Observed by AFM and Multiplication Behaviors 62
4.4.1 Motivation 62
4.4.2 Morphology of Au Film on 1, 4, 5, 8-Naphthalenetetracarboxylic Dianhydride (NTCDA) [16] 62
4.4.3 Revised Structural Trap Model 63
4.4.4 Three-Dimensional Energy Structure 64
4.4.5 Effects of Metal Nanoparticle Size 65
4.4.6 Rear Surface of Au and In Films 67
4.5 Essential Factors in the Structural Trap Model 69
4.5.1 Spatial Gap 69
4.5.2 Molecular-Sized Roughness 69
4.5.3 True Nature of Molecular-Sized Roughness 70
4.6 Conclusion 70
References 70

5 Photocurrent Multiplication in Organic Single Crystals—Molecular Blind Alleys 73
5.1 Background 73
5.2 Motivation 73
5.3 Multiplication Behaviors of Organic Single Crystals 74
5.3.1 Naphthalene Tetracarboxylic Anhydride (NTCDA) 74
5.3.2 Single-Crystal Growth 74
5.3.3 Cell Fabrication 75
5.3.4 Multiplication Rate 76
5.3.5 Spectral Sensitivity 77
5.3.6 Response of Multiplied Photocurrent 78
5.3.7 Multiplication Mechanism 79
5.4 Real Nature of Structural Trap 81
5.4.1 Anomalous Characters of Traps at Organic/Metal Interface 81
5.4.2 Structural Trap Model Hypothesis 82
5.4.3 The Idea Behind the Hypothesis 83
5.4.4 Physically Existing Nanostructure Acting as Structural Trap 83
5.4.5 Spatial Gap 83
5.4.6 Molecular-Size Roughness 84
5.5 Molecular Blind Alley Model 84
5.5.1 Why Single Crystal? 84
5.5.2 Molecular Steps 85
5.5.3 Molecular Steps Acting as Molecular Blind Alleys 86
5.5.4 Molecular Step Density Versus Multiplication Rate 87
5.5.5 Molecular Kinks Acting as Molecular Blind Alley Sites 88
5.5.6 Deliberate Formation of Molecular Blind Alleys 90
5.5.7 Molecular Blind Alleys in Polycrystalline NTCDA Films 90
5.6 Future Perspective 92
5.6.1 High Resolved AFM Observation of Steps and Kinks 92
5.6.2 Multiplication Control via the Design of Molecular Blind Alleys 92
5.6.3 Design of Molecular Blind Alleys by Nanoimprint Lithography 94
5.7 Conclusion 94
References 94

6 Photocurrent Multiplication at Organic Heterojunctions 97
6.1 Background 97
6.2 Motivation 97

6.3 Photocurrent Multiplication at CuPc/Me-PTC Heterojunction ... 98
6.3.1 CuPc/Me-PTC Heterojunction ... 98
6.3.2 Multiplication Rate ... 99
6.3.3 Identification of Interface ... 99
6.3.4 Multiplication Suppression by Superimposed Light ... 100
6.3.5 Multiplication Mechanism ... 100
6.3.6 Suppression Mechanism ... 101
6.3.7 Structural Trap at Organic Heterojunction ... 103
6.4 Organic Heterojunctions Incorporating Hole and Electron Transporting Layers ... 103
6.4.1 Multiplication Rate ... 104
6.4.2 Action Spectra ... 105
6.4.3 Multiplication Mechanism ... 106
6.4.4 Structural Traps at HTL/*n*-OSC and ETL/*p*-OSC HTL (ETL)/*n*-OSC (*p*-OSC) Heterojunctions ... 108
6.5 Conclusion ... 109
References ... 109

7 High-Speed Response Devices ... 111
7.1 Background ... 111
7.1.1 Transient Response of Photocurrent Multiplication ... 111
7.1.2 Structural Trap Model ... 111
7.2 Numerical Calculation ... 113
7.2.1 Motivation ... 113
7.2.2 Simplified Model for Calculation ... 113
7.2.3 Time Development of Hole Accumulation ... 114
7.2.4 Surface Mobility ... 115
7.2.5 Keys of High-Speed Response ... 116
7.3 High-Speed Response Devices Having C_{60}:CuPc Co-deposited Films ... 117
7.3.1 Motivation ... 117
7.3.2 Cell Structure ... 117
7.3.3 Photoresponse of Multiplied Photocurrent ... 117
7.3.4 Factors of High-Speed Response ... 120
7.4 High-Speed Devices Having Double-Layered Structure ... 121
7.4.1 Motivation ... 121
7.4.2 Double-Layered Structure ... 121
7.4.3 Transient Response ... 122
7.4.4 Film Morphology ... 122
7.4.5 Dark Current Suppression ... 123
7.4.6 Multiplication Rate Versus Applied Voltage ... 124
7.4.7 Transient Response Versus Applied Voltage ... 124
7.4.8 Factors of High-Speed Response ... 125

7.5 Requirements of Organic Multiplication-Type Photosensors ... 125
7.6 Conclusion ... 126
References ... 126

8 Effect of Oxygen and Water on Photocurrent Multiplication Rates ... 129
8.1 Background ... 129
8.2 Motivation ... 129
8.3 Effect of O_2 Under Ex-Situ Conditions ... 130
8.3.1 Ex-Situ Conditions ... 130
8.3.2 *p*- and *n*-Type OSC Cells ... 130
8.3.3 *p*-DQ Cells ... 131
8.3.4 *n*-NTCDA Cells ... 131
8.3.5 O_2 Enhancement Mechanism at *p*-DQ/Au Junction ... 132
8.3.6 O_2 Suppression Mechanism at *n*-NTCDA/Au Junction ... 134
8.3.7 A Concept of Multiplied Gas Detection ... 135
8.4 Effects of H_2O on Photocurrent Multiplication Observed Under In-Situ Conditions ... 135
8.4.1 Irreversible Change Due to H_2O Adsorption ... 135
8.4.2 Motivation ... 136
8.4.3 In-Situ Conditions ... 136
8.4.4 Cells ... 136
8.4.5 Effect of O_2 Under In-Situ Conditions ... 137
8.4.6 Effect of H_2O ... 138
8.4.7 Enhancement/Suppression Mechanisms of O_2 Under In-Situ Conditions ... 139
8.4.8 Multiplication Rate Enhancement Mechanism of H_2O ... 140
8.4.9 Future Work ... 140
8.5 Conclusion ... 141
References ... 141

9 Multiplied Photocurrent Oscillation with Negative Resistance ... 143
9.1 Motivation ... 143
9.2 Cell Structure and Measurement Conditions ... 143
9.3 Photocurrent Oscillation and Voltage-Controlled Measurements ... 144
9.4 S-shaped Negative Resistance: Current-Controlled Measurements ... 145
9.5 Negative Resistance and Heterogeneity ... 145
9.6 Thickness Dependence ... 146
9.6.1 H_2Pc Thickness ... 146
9.6.2 NTCDA Thickness ... 148
9.7 Heterogeneous Accumulation Model ... 150
9.8 Feedback Effect ... 154

9.9 Generality ... 154
9.10 Removal of H_2Pc Roughness ... 155
9.10.1 Mechanical Pressing ... 155
9.10.2 H_2Pc Roughness ... 156
9.10.3 Oscillation Behavior and H_2Pc Roughness ... 156
9.10.4 Oscillation Disappearance ... 159
9.10.5 Conclusion ... 159
References ... 160

10 Avalanche Multiplication in Perylene Molecular Crystals ... 163
10.1 Background ... 163
10.2 Motivation ... 163
10.3 Photocurrent Multiplication Rate in Perylene Single Crystals ... 165
10.3.1 Perylene Single Crystals ... 165
10.3.2 Surface-Type Cells ... 165
10.3.3 Dark Current ... 166
10.3.4 Multiplication Characteristics ... 166
10.3.5 Photoresponse ... 168
10.3.6 Carrier-Traveling Distance ... 169
10.4 Impact Ionization ... 169
10.5 Ionization Rate ... 171
10.6 Mean Free Path ... 171
10.7 Conclusion ... 172
References ... 172

11 Progress in Organic Photocurrent Multiplication ... 175
11.1 Aim of This Chapter ... 175
11.2 Historical Outline ... 175
11.3 Blended Junction ... 178
11.4 Blocking Layer ... 178
11.4.1 Blocking Layer at Active Interface ... 178
11.4.2 Blocking Layer at Counter Interface ... 180
11.5 Carrier Trap ... 181
11.5.1 Uniform Trap ... 181
11.5.2 Interfacial Traps ... 181
11.5.3 Incomplete Percolation Traps ... 183
11.6 Universality of Multiplication Concept—Hybrid System ... 184
11.7 Multiplication in Charge Transfer Region ... 186
11.8 Essence of Progress ... 188
11.9 Organic Photocurrent Multiplication (OPM) Devices ... 188
11.10 Conclusion ... 189
References ... 189

12 Perspective on Organic Photocurrent Multiplication 191
12.1 Aim of This Chapter 191
12.2 Artificial Fabrication 191
12.2.1 Electrode Metal Having the Tip of the Needle Shape 191
12.2.2 Molecular Blind Alley Traps Made by Nanoimprint Lithography 192
12.2.3 Micrometer-Sized Thickness Heterogeneity Made by Nanoimprint Lithography 193
12.3 Direct Design of Molecular Blind Alley 194
12.3.1 Detailed Observation of Steps and Kinks 194
12.3.2 Control of Multiplication Behaviors by Designing Steps and Kinks 195
12.4 Insulating Layer 196
12.4.1 Metal/Insulator/Semiconductor (MIS) Junction 196
12.4.2 Spatial Separation 196
12.4.3 How to Break the Tunneling Balance Between Electrons and Holes? 197
12.5 Organic Avalanche Multiplication 198
12.6 Conclusion 198
References 200

Chapter 1
Photocurrent Multiplication in Amorphous Silicon Carbide Films

1.1 Photocurrent Multiplication in Amorphous Silicon Carbide Films

Photocurrent multiplication was discovered in amorphous silicon carbide (a-SiC:H) films in the sandwich-type cell structure of SnO_2/a-SiC:H/Au. The multiplication rate was 300 times. This phenomenon, which appeared only under the irradiation of bulk-absorbed red light, is not due to avalanche effects, but is thought to be due to electron tunneling from SnO_2 to the a-SiC:H film through a thin insulating silicon oxide layer formed at the a-SiC:H/SnO_2 interface. This occurred under a high electric field built up by photo-accumulated space charges (trapped holes) near the interface.

1.1.1 Background

Amorphous silicon-based films have been widely used in various optoelectronic devices such as solar cells [1–3], photoreceptors in electrophotography [4, 5], and thin-film transistors in liquid–crystal displays [6, 7]. In contrast to crystalline materials, photocurrent multiplication phenomena have not yet been reported in amorphous Si-based films, with the exception of avalanche multiplication in amorphous selenium [8]. Some attempts have been made to achieve photocurrent multiplication in these films with rather complicated structures of the superlattice [9] or the *p-i-n* junction diode, with an insulating amorphous silicon nitride (a-SiN:H) thin film for the *i* layer [10]. Thus, to that point in time, there had been few reports of photocurrent multiplication for amorphous silicon-based semiconductors.

M. Hiramoto, *Organic Photocurrent Multiplication*, Electronic Materials: Science & Technology, https://doi.org/10.1007/978-981-99-1237-7_1

1.1.2 Motivation

In 1998, Tang reported a famous study on organic electroluminescent diodes [11, 12]. Inspired by this work, we fabricated a new type of light transducer consisting of an organic electroluminescent diode on a photoconductive amorphous silicon carbide (a-SiC:H) as an electron photoinjecting electrode [13], which enables light up-conversion from red to green [14]. During these studies, the author realized that a large photocurrent multiplication reaching several factors of ten had occurred in the a-SiC:H films. Therefore, the author began studying single a-SiC:H films sandwiched between metal electrodes, and discovered the photocurrent multiplication phenomenon [15].

1.1.3 Discovery of Photocurrent Multiplication

In Fig. 1.1, the cell structure where the photocurrent multiplication phenomenon was first observed is shown. The 1 μm-thick a-SiC:H films were deposited by plasma chemical vapor deposition (CVD) (100 W) on SnO_2 (NESA) glass substrates maintained at 250 °C under the gas flow of a mixture of SiH_4 (20 cm^3s^{-1}), C_2H_2 (1 cm^3s^{-1}), and H_2 (700 cm^3s^{-1}) [5]. PH_3 gas (5 ppm) was added to the mixture to produce *n*-type films. Finally, a semitransparent Au electrode (20 nm) was deposited onto the a-SiC:H films. Photocurrent measurements were carried out in air using an electrometer (Keithley 485) with an applied voltage between the Au and SnO_2 electrodes.

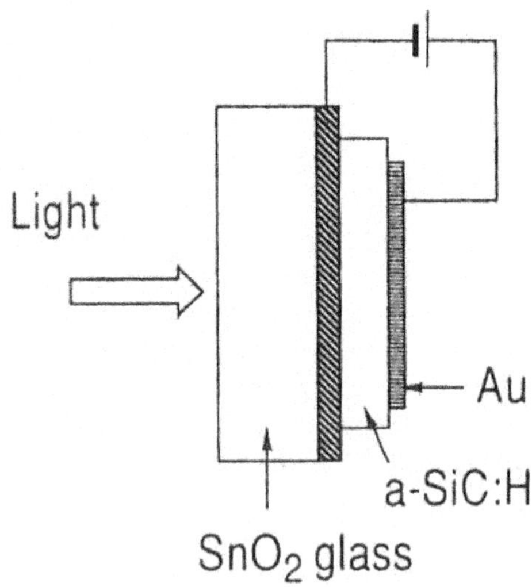

Fig. 1.1 Device structure. Reproduced with the permission of [15]. © 1991 American Institute of Physics

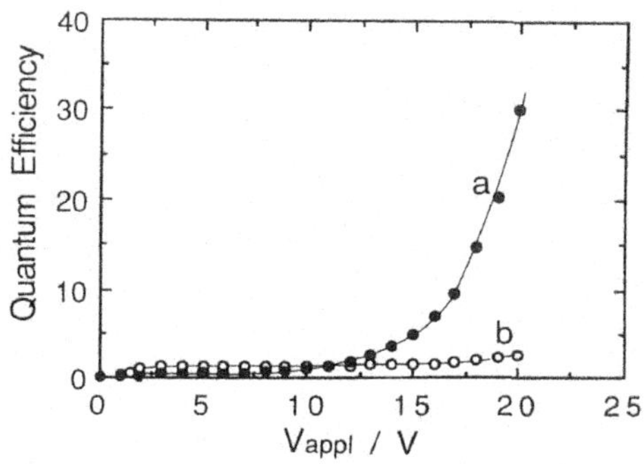

Fig. 1.2 Dependence of the photocurrent quantum efficiency on applied voltage (V_{appl}) under **a** He–Ne laser (632.8 nm) and **b** Ar-ion laser (488 nm) irradiations through negatively biased SnO_2 electrode. Light intensity was 5 $mWcm^{-2}$. Reproduced with the permission of [15]. © 1991 American Institute of Physics

Figure 1.2 shows the typical applied voltage (V_{appl}) dependences of the photocurrent quantum efficiency under the irradiation of He–Ne (632.8 nm) (curve a) and Ar ion (488 nm) (curve b) laser lights, respectively. The photocurrent quantum efficiency, that is, the multiplication rate, is defined as the ratio of the number of photogenerated carriers collected to the number of photons absorbed by the a-SiC:H film. Each of the two light sources caused a different behavior when a negative voltage was applied to the illuminating SnO_2 electrode. Under blue Ar-ion laser irradiation (curve b), the photocurrent quantum efficiency reached unity near 2 V and remained constant with increasing V_{appl}. However, when irradiated by the red He–Ne laser (curve a), surprisingly, the photocurrent quantum efficiency increased sharply above 10 V and reached several multiples of ten at 20 V. The data shown here are for the case where a negative SnO_2 electrode was irradiated, but a similar phenomenon was observed for the case of illuminating red light through a semitransparent Au electrode. Multiplication was also observed when the Au electrode was negative, but a much larger V_{appl} was necessary. The results presented here are for conditions in which a negative voltage is applied to the illuminated SnO_2 electrode.

1.1.4 Spectral Dependence

The spectral dependence of the photocurrent quantum efficiency is shown in Fig. 1.3, together with the absorption spectrum of the 1 μm a-SiC:H film used in this study. It is evident that photocurrent multiplication occurs only in the spectral region with wavelengths longer than 550 nm, where the a-SiC:H film has a small absorption coefficient that allows bulk excitation of the film. By contrast, in the region of short wavelengths where the excitation is confined near the film surface, the photocurrent quantum efficiency remains almost unity. This result indicates that the present

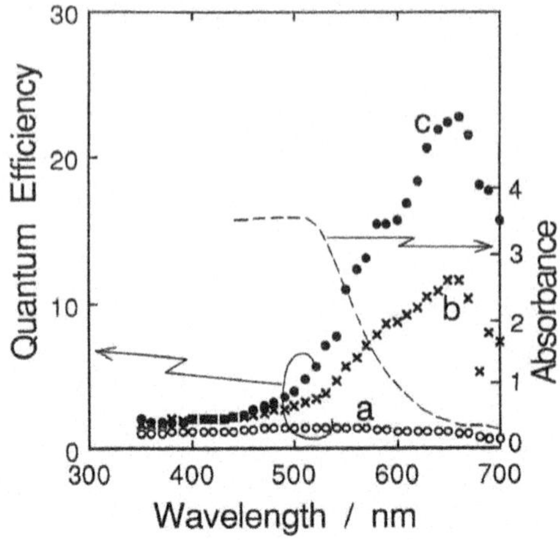

Fig. 1.3 Spectral dependence of photocurrent quantum efficiency at **a** 5 V, **b** 15 V, and **c** 20 V under the irradiation of monochromatic light (5 $mWcm^{-2}$) through SnO_2, electrode. Broken curve represents the absorption spectrum of 1 μm-thick a-SiC:H film. Reproduced with the permission of [15]. © 1991 American Institute of Physics

photocurrent multiplication cannot be explained by the avalanche effect, in which the photocurrent multiplication is independent of the incident light wavelength.

1.1.5 Light Intensity Dependence

Figure 1.4 shows the dependence of the photocurrent density I_{ph} on the input light intensity L [15, 16]. Under the condition that no multiplication takes place (V_{appl} = 5 V), the photocurrent is proportional to the light intensity (L) (Fig. 1.4, line a). However, the excess photocurrent due to the multiplication (V_{appl} = 20 V) exhibited a dependence of $L^{0.5}$ (Fig. 1.4, line b), implying the existence of a recombination-limited process in the multiplied photocurrent generation. At relatively low light intensities (50 μWcm^{-2}), the multiplication rate reached 300.

1.1.6 Transient Response

The multiplied photocurrent under weak red-light irradiation exhibited a transient response on the order of several tens of seconds to switching either light or voltage, as shown in Fig. 1.5 [15, 16]. The response time depended on the light intensity, that is, lower intensity resulted in slower rise and decay curves for light on and off. The response in Fig. 1.5 is for the case of low light intensities of 21.0 and 5.3 μWcm^{-2}. This suggests that the multiplication process involves trapping of the photogenerated carriers.

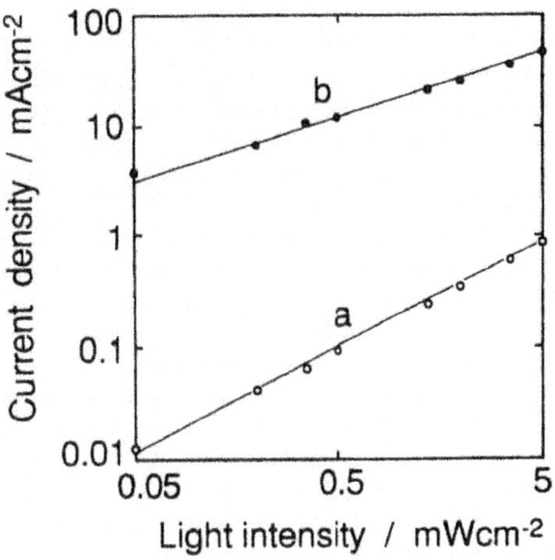

Fig. 1.4 Light intensity versus photocurrent density. $V_{appl} = 5$ V (line a) and 20 V (line b). The He–Ne laser was irradiated from the SnO_2 side

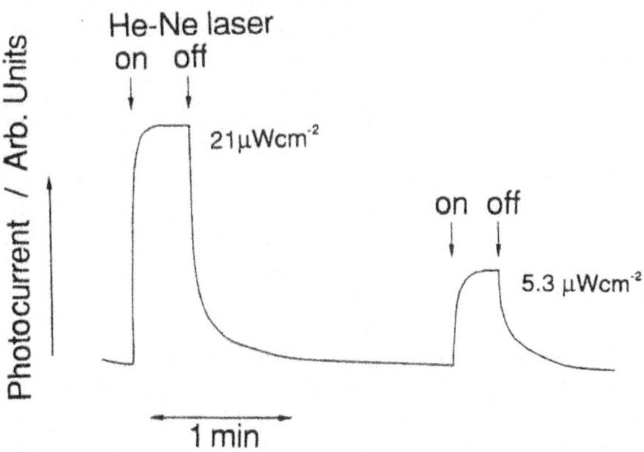

Fig. 1.5 Photocurrent response for switching the He–Ne laser light, of which intensities are 21 and 5.3 μWcm^{-2} at $V_{appl} = 20$ V. Photocurrent quantum efficiency, i.e., the multiplication rate, exceeded 60. Reproduced with the permission of [15].

1.1.7 Interfacial Structure

Because of the peculiar spectral response of the multiplied photocurrent, the author suspected that the key processes of the observed photocurrent multiplication occurred at the interface between the electrode and a-SiC:H. Thus, the depth profile of the a-SiC:H (50 nm)/SnO_2 film was analyzed by X-ray photoelectron spectroscopy (XPS) with the ion-etching technique [15, 16]. The dependence of the atomic ratio and photoelectron spectra of the Si_{2p} peak on the Ar ion etching time are shown in Fig. 1.6a, b, respectively. The etching time from 0 to 220 s corresponds to the a-SiC:H layer, and that after 220 s corresponds to the SnO_2 film. The indistinct change in the atomic ratio at the a-SiC:H/SnO_2 interface was caused by the rough surface of the 50 nm-thick SnO_2 film. The photoelectron spectra of Si near the free surface of the a-SiC:H film from 0 to 30 s (Fig. 1.6b) was accompanied by a higher binding energy component at approximately 104 eV, which can be assigned to silicon oxide

(SiO_x). This component disappeared in the bulk of a-SiC:H at 60 s, but appeared again at the a-SiC:H/SnO_2 interface after 220 s, that is, 360 and 420 s in Fig. 1.6b. This result suggests the existence of a thin SiO_x layer between the a-SiC:H and SnO_2. The thickness of these SiO_x layers formed at the free surface and the a-SiC:H/SnO_2 interface was estimated to be less than several nanometers.

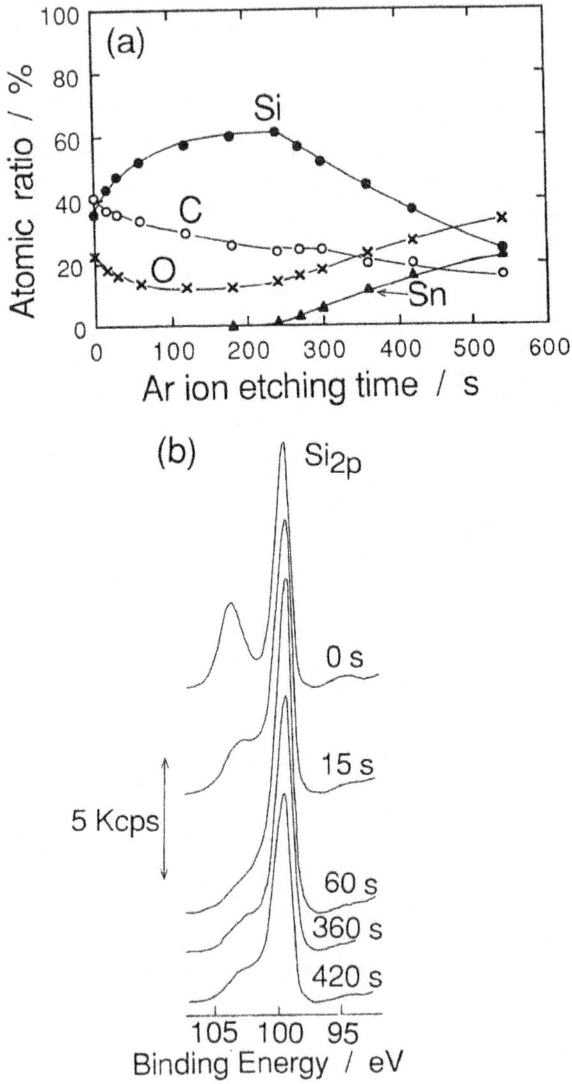

Fig. 1.6 **a** Atomic ratio for the layered film having a structure of a-SiC:H (50 nm)/SnO_2 as a function of Ar ion etching time. **b** Photoelectron spectra of Si_{2p} peak after Ar etching time for 0, 15, 60, 360, and 420 s

1.1.8 Multiplication Mechanism

Because the present multiplication phenomenon depends on the incident light wavelength, it cannot be explained by the avalanche effect. A very thin SiO_x layer exists at the SnO_2/a-SiC:H and a-SiC:H/Au interfaces. The multiplied photocurrent exhibited a light intensity (L) dependence of $L^{0.5}$, which is different from that of the normal photocurrent ($L^{1.0}$) This fact implies that some recombination-limited processes should be considered in the mechanism. Moreover, the observed transient response suggests that the multiplication process includes trapping of photogenerated carriers.

The following mechanism was proposed assuming that interfacial states exist at the SiO_x/SiC:H interface, as illustrated schematically in Fig. 1.7. Irradiation of the bulk absorbed red light generates electrons and holes drifting along the potential gradient (a), and the holes reaching the a-SiC:H/SnO_2 interface are partly trapped by the interfacial states (b). Accumulation of trapped holes leads to a high electric field, which is distributed across a thin insulating SiO_x film, and finally gives rise to the tunneling injection of electrons through the thin SiO_x layer from the SnO_2 electrode to the conduction band of a-SiC:H (c).

This model explains the transient response of the multiplied photocurrent as a phenomenon related to hole trapping and emission processes. Moreover, the $L^{0.5}$ dependence of the multiplied photocurrent on light intensity reflects the recombination process between trapped holes at the interface and the photogenerated electrons, because the number of electrons near the interfacial states increases with light intensity. The reduced photocurrent multiplication in the shorter-wavelength region is interpreted to be caused by the efficient recombination between trapped holes and electrons photogenerated near the surface at high excitation density, which decreases the number of accumulated holes trapped at the interface.

Because there is also a thin SiO_x layer at the a-SiC:H/Au interface, the multiplication phenomenon in the case of negative voltage applied to the Au electrode can be considered to be caused by the same mechanism at the a-SiC:H/SnO_2 interface.

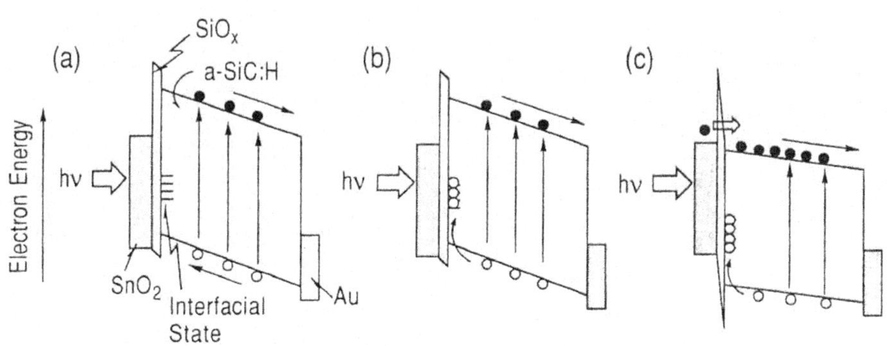

Fig. 1.7 Energy structures of an SnO_2/a-SiC:H/Au cell under applying negative voltage to SnO_2 electrode. **a** Red-light excitation. **b** Hole trapping by interfacial states. **c** Electron tunneling through an SiO_x layer. Reproduced with the permission of [15]. © 1991 American Institute of Physics

In this case, however, a much higher applied voltage is required for effective electron injection from the Au electrode to the conduction band of a-SiC:H, presumably because of the larger work function of Au compared to that of SnO_2. Schmidlin [17, 18] proposed a general amplification model, taking the interfacial levels into account, and interpreted the photocurrent gain observed in cells consisting of a photoconductor and a thin insulator. The photocurrent multiplication in our simple cell seems to occur as a result of the unintentional formation of a thin insulating layer at the interface. Nakayama et al. [19–21] proposed a different multiplication model, involving the electron tunneling process through the Schottky barriers formed at the interfaces of SnO_2/a-SiC:H and Au/a-SiC:H. At this stage, we thought that further study was necessary for the strict identification of the insulating layer.

1.1.9 Temperature Dependence

The magnitude of the multiplied photocurrent should be independent of temperature because the tunneling process is independent of temperature. Figure 1.8 shows the temperature dependence of the photocurrent quantum efficiency [16]. At temperatures below −50 °C, the quantum efficiency was almost constant, as expected. However, above −50 °C, it decreased with increasing temperature. This behavior can be explained by the thermal detrapping of trapped holes at the interfacial states, which reduces the electric field across the insulating layer.

1.2 Photo-Modulation of Photocurrent Multiplication

The photocurrent multiplication observed under the irradiation of bulk-absorbed red light in amorphous silicon carbide (a-SiC:H) films could be photo-modulated

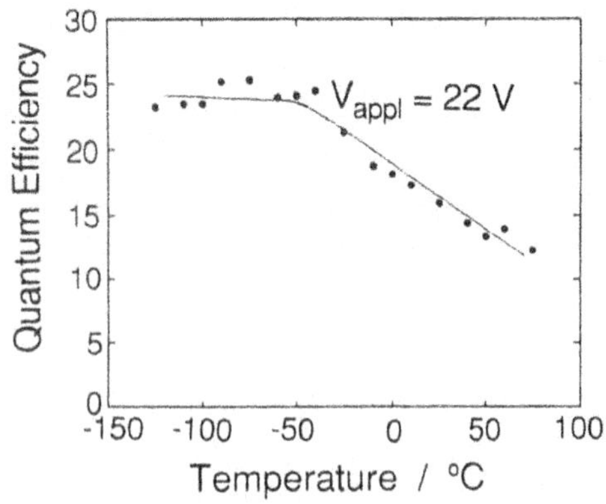

Fig. 1.8 Temperature dependence of the photocurrent quantum efficiency at $V_{appl} = 22$ V under the irradiation of He–Ne laser (3 $mWcm^{-2}$)

depressively by superimposing the surface-absorbed blue light on the irradiated red light. This phenomenon strongly supports the proposed multiplication mechanism involving the accumulation of trapped holes near the SnO_x/a-SiC:H interface, because the multiplication suppression can be reasonably explained by the disappearance of accumulated holes due to recombination with electrons photo-generated by the surface-absorbed blue light [22].

1.2.1 Basic Idea

The photocurrent multiplication occurs by electron tunneling from SnO_2 to an a-SiC:H film through a thin insulating SiO_x layer under a high electric field created by the photo-accumulated trapped holes near the SiO_x/a-SiC:H interface (Fig. 1.7). Interestingly, this phenomenon appeared only under irradiation with bulk-absorbed red light. The author envisioned that in the region with wavelengths shorter than 500 nm (Fig. 1.3), the multiplication process is suppressed by irradiation with shorter wavelength light, which is strongly absorbed by the a-SiC:H film with a large absorption coefficient of more than 3 (Fig. 1.3, broken curve). Thus, the author proposed that the multiplied photocurrent generated by bulk absorbed wavelengths longer than 500 nm might be suppressed by short-wavelength light superimposed on the red light.

1.2.2 Multiplication Suppression

The cell used in this study had the same structure as that shown in Fig. 1.1. The a-SiC:H film had the same thickness of 1 μm. A 670 nm diode laser was used as the red light source. As a superimposed light source, monochromatic light from a 500 W xenon lamp through a monochromator was used (Fig. 1.9 (upper)) [22]. As previously reported in Sect. 1.1. [14, 15], large photocurrent multiplication occurred under the irradiation of bulk-absorbed red light through the negatively biased SnO_2 electrode (V_{appl} > 10 V) with a photocurrent quantum efficiency (multiplication rate) of several tens. As clearly demonstrated in Fig. 1.9 (lower), under the condition of photocurrent multiplication, the superimposed irradiation of surface-absorbed UV light (400 nm) effectively suppressed the excess photocurrent, that is, decreased the multiplication factor. This response was reversible. The suppression rate (S_r) of the photocurrent induced by UV irradiation was strongly dependent on the temperature as well as the intensity ratio of the two wavelength lights (I_{UV}/I_{red}). Figure 1.10 shows S_r at −100 and 25 °C as a function of I_{UV}/I_{red}. The maximum S_r was approximately 20% at room temperature; however, it increased to 80% at −100 °C. S_r was not linearly proportional to the I_{UV}/I_{red}. It increased sharply below $I_{UV}/I_{red} = 0.2$ and was almost saturated above $I_{UV}/I_{red} = 0.4$. These results suggest that photocurrent multiplication can be intentionally photo-controlled over a wide range by small modulations of the superimposed UV-light intensity.

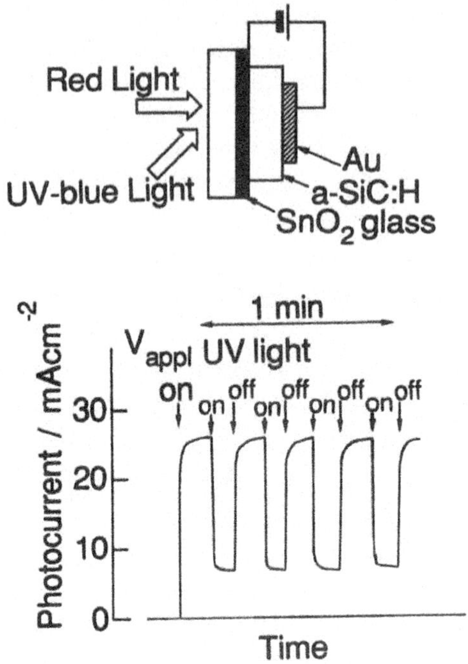

Fig. 1.9 The device structure and typical photocurrent response on UV irradiation at −100 °C. Red light from a diode laser (670 nm, 3 mWcm^{-2}) was always irradiated. Applied voltage (V_{appl}) was 20 V. The UV light (400 nm) was superimposed with the intensity of 6.8 mWcm^{-2}. Reproduced with the permission of [22]. © 1991 American Institute of Physics

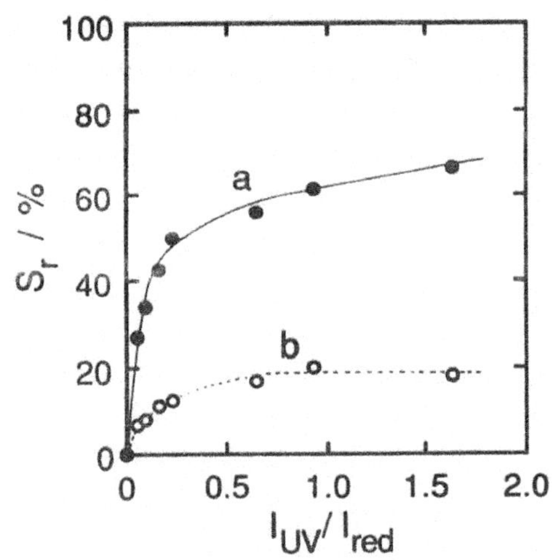

Fig. 1.10 The suppression rate (S_r) of the photocurrent multiplication as a function of the intensity ratio of two wavelength lights (I_{uv}/I_{red}). Experiments were carried out at −100 (**a**) and 25 °C (**b**). The intensity of 400 nm UV light (I_{uv}) is 0.36 mWcm^{-2}. Reproduced with the permission of [22]. © 1991 American Institute of Physics

1.2.3 Suppression Wavelength Region

Figure 1.11 shows the excess photocurrent density under the simultaneous irradiation of 670 nm red light from a diode laser and monochromatic light of a wavelength indicated on the lower horizontal axis [22]. The broken line in Fig. 1.11b denotes

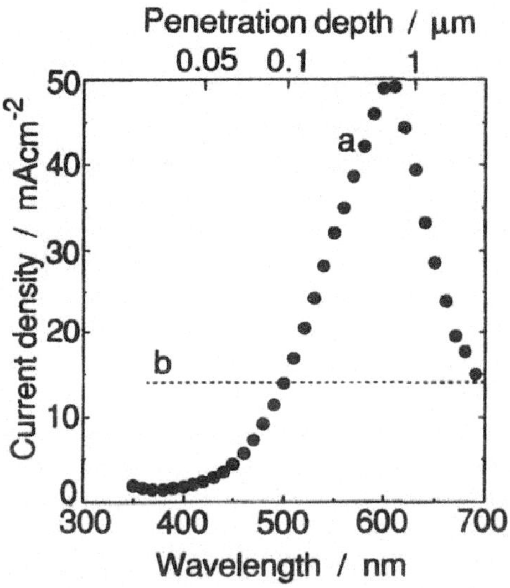

Fig. 1.11 Photocurrent density (curve a) at $V_{appl} = 15$ V under the simultaneous irradiation of red light from a diode laser (3 $mWcm^{-2}$) and respective monochromatic light (4 $mWcm^{-2}$) of wavelengths shown by lower horizontal axis. A broken line **b** shows the photocurrent density under the irradiation of diode laser (670 nm) only. Light penetration depth for the present a-SiC:H film is shown on the upper horizontal axis. Measurements were carried out at −100 °C. Reproduced with the permission of [22]. © 1991 American Institute of Physics

the photocurrent density under the irradiation of the diode laser light only, which was fixed to a quantum efficiency of approximately 10 throughout the experiments. The multiplication was effectively suppressed by the superimposed monochromatic light wavelengths shorter than 480 nm (Fig. 1.11, curve a). The penetration depth of the respective light wavelength, which means that 90% of the incident photons are absorbed within this thickness of the a-SiC:H film, is shown on the upper horizontal axis. Thus, this result clearly proves that the high excitation density near the SnO_x/a-SiC:H interface by the surface-absorbed ultraviolet-blue (350–500 nm) monochromatic light causes drastic suppression of the photocurrent multiplication.

1.2.4 Suppression Mechanism

The suppression mechanism is illustrated in Fig. 1.12 [22]. In the previous section, a mechanism for photocurrent multiplication under irradiation by bulk-absorbed red light was proposed (Fig. 1.7). When the red light is irradiated with a red SnO_2 electrode, holes photogenerated by bulk excitation drift toward the negative SnO_2 electrode, and some of them are trapped by the interfacial states between a-SiC:H and SiO_x. Accumulation of trapped holes leads to a high electric field distributed across an insulating SiO_x film, which finally induces the tunneling injection of electrons

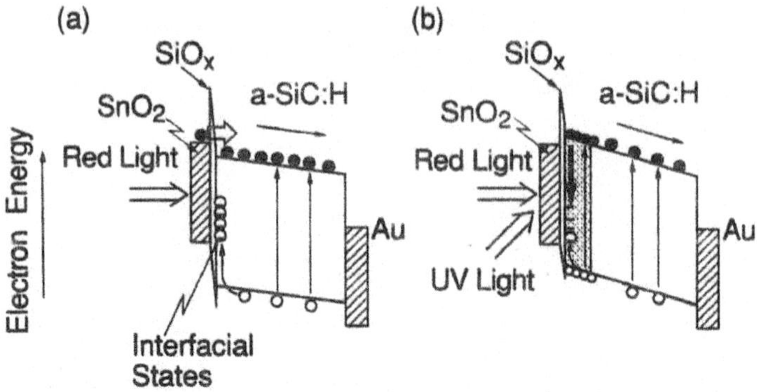

Fig. 1.12 The energy structure and schematic illustration for the photocurrent multiplication in SnO_2/a-SiC:H/Au cell under applying the negative voltage to SnO_2 electrode (Fig. 1.9) (upper). **a** Photocurrent multiplication under the irradiation of bulk-absorbed red light. **b** Suppression of the photocurrent multiplication by the irradiation of surface-absorbed (shaded area) UV-blue light superimposed on the red light. Reproduced with the permission of [22]. © 1991 American Institute of Physics

from the SnO_2 electrode to a-SiC:H (Fig. 1.12a). The irradiation of surface-absorbed UV-blue light superimposed on red light generates electrons with high density in the region close to the SnO_2/a-SiC:H interface (Fig. 1.12b, shaded area). These photogenerated electrons dissipate the holes trapped at the interface by recombination, resulting in an effective reduction of the electric field across the SnO_2/a-SiC:H interface. As a result, effective suppression of the tunneling injection of electrons from the SnO_2 electrode occurs. It should be noted that the observed photocurrent suppression by UV light itself strongly supports the previously proposed mechanism based on electron tunneling for photocurrent multiplication (Fig. 1.7). Moreover, the above results successfully demonstrate that one can photocontrol the photocurrent multiplication can be photocontrolled in a-SiC:H films over a wide range, as desired, by choosing factors such as the applied voltage, temperature, intensity ratio of UV light to red light, and penetration depth of UV-blue light.

1.3 Conclusion

Photocurrent multiplication that reached a value of 300 was discovered in a simple cell with SnO_2/a-SiC:H/Au structure under the irradiation of red light. A mechanism was based on photoinduced electron tunneling injection from a SnO_2 electrode to an a-SiC:H film through a thin SiO_x layer formed at the a-SiC:H/SnO_2 interface, which requires the accumulation of photoinduced holes near the interface to create high electric fields across the insulating SiO_x layer. This model provides a reasonable explanation for the observed dependence of the multiplication rate on the light wavelength, light intensity, and temperature. Moreover, photocontrol of the photocurrent multiplication by superimposing UV-blue light on red light was demonstrated. This further proves the proposed multiplication mechanism. The present photocurrent multiplication has the potential to develop highly sensitive a-SiC:H-based image sensors.

Acknowledgements The authors thank Professor T. Kawamura and Dr. Y. Nakayama of the University of Osaka Prefecture for supplying the a-SiC:H samples. The author is grateful to Prof. H. Tsubomura of Osaka University for his kind accommodation of the XPS measurements.

References

1. Spear, W.E., Lecomber, P.E.: Substitutional doping of amorphous silicon. Solid State Commun. **17**, 1193–1196 (1975)
2. Carlson, D.E.: Amorphous silicon solar cells. IEEE Trans. Electron Devices ED **24**, 449–453 (1977)
3. Tawada, Y., Yamaguchi, T., Nonomura, S., Hotta, S., Okamoto, H., Hamakawa, Y.: Optimizations of the film deposition parameters for the hydrogenated amorphous silicon solar cell. Jpn. J. Appl. Phys. **20**(Suppl. 20–2), 219–225 (1981)
4. Nakayama, Y., Sugimura, A., Nakano, M., Kawamura, T.: Photogr. Sci. Eng. **26**, 188 (1982)
5. Nakayama, Y., Akita, S., Wakita, K., Kawamura, T.: New types of electrophotographic photoreceptors using photosensitive a-SiC: H and a-Si:H. Mater. Res. Soc. Symp. Proc. **118–423**, 73 (1988)
6. LeComber, P.G., Spear, W.E., Dhaith, A.: Electron. Lett. **15**, 179 (1979)
7. Snell, A.J., Mackenzie, K.D., Spear, W.E., LeComber, P.G., Hughes, A.: Application of amorphous silicon field effect transistors in addressable liquid crystal display panels. J. Appl. Phys. **24**, 357–362 (1981)
8. Takasaki, Y., Tsuji, K., Hirai, T., Maruyama, E., Tanioka, K., Yamazaki, J., Shidara, K., Taketoshi, K.: Avalanche multiplication of photo-generated carriers in amorphous semiconductor, and its application to imaging device. Mater. Res. Symp. Proc. **118**, 387 (1988)
9. Jwo, S.-C., Wu, M.-T., Fang, Y.-K., Chen, Y.-W., Hong, J.-W., Chang, C.-Y.: Amorphous silicon/silicon carbide superlattice avalanche photodiodes. IEEE Trans. Electron. Dev. **35**, 1279–1283 (1988)
10. Yoshimi, M., Hattori, K., Okamoto, H., Hamakawa, Y.: Spring Meeting of Material Research Society, Symposium Proceedings, Amorphous Silicon Technology, San Francisco (1990)
11. Tang, C.W., VanSlyke, S.A.: Organic electroluminescent diodes. Appl. Phys. Lett. **51**, 913–915 (1987)
12. Tang, C.W., VanSlyke, S.A., Chen, C.H.: Electroluminescence of doped organic thin-films. J. Appl. Phys. **65**, 3610–3616 (1989)
13. Hiramoto, M., Miyao, T., Yokoyama, M.: Spatially addressable light transducer using an organic electroluminescent diode combined with amorphous silicon carbide film as an electron photoinjecting electrode. Appl. Phys. Lett. **57**, 1625–1627 (1990)
14. Hiramoto, M., Yoshimura, K., Miyao, T., Yokoyama, M.: Up-conversion of red light to green by a new type of light transducer using organic electroluminescent diode combined with photoresponsive amorphous silicon carbide. Appl. Phys, Lett. **58**, 1146–1148 (1991)
15. Hiramoto, M., Yoshimura, K., Nakayama, Y., Akita, S., Kawamura, T., Miyao, T., Yokoyama, M.: Photocurrent multiplication in amorphous silicon carbide films. Appl. Phys. Lett. **59**, 1992–1994 (1991)
16. Hiramoto, M., Yoshimura, K., Yokoyama, M.: Photocurrent multiplication in amorphous silicon carbide films. J. Imag. Sci. Technol. **37**, 192–196 (1993)

17. Schmidlin, F.W., Tutihasi, S.: Amorphous and liquid semiconductors. In: Proceedings of 5th International Conference, 3–8 Sept 1973, vol. 1, p. 381 (1974)
18. Schmidlin, F.W.: Electrical switching device based on charge-controlled double injection. Phys. Rev. B **1**, 1583–1587 (1970)
19. Nakayarna, Y., Yajima, T., Ueda, H., Akita, S., Kawamura, T.: In: Proceedings of the 14th International Conference on Amorphous Semiconductors, Garmisch-Partenkirchen, Germany, 19–23 Aug 1991
20. Nakayarna, Y., Yajima, T., Ueda, H., Akita, S., Kawamura, T.: Two-way photocurrent-multiplication in amorphous silicon carbide cells. J. Non-cryst. Solids **137&138**, 1279–1282 (1991)
21. Akita, S., Ueda, H., Nakayarna, Y.: Mechanism of photocurrent multiplication in amorphous silicon carbide Schottky cells. J. Appl. Phys. **77**, 1120–1125 (1995)
22. Hiramoto, M., Yoshimura, K., Yokoyama, M.: Photomodulation of photocurrent multiplication in a high gain amorphous silicon carbide film. Appl. Phys. Lett. **60**, 1102–1104 (1992)

Chapter 2
Photocurrent Multiplication in Organic Semiconductor Films

2.1 Discovery

2.1.1 An Original Idea: Taking Out the Interface

In 1991, the phenomenon of photocurrent multiplication was discovered in amorphous silicon carbide (a-SiC:H) films, as described in Chap. 1 [1–3]. The multiplication process occurs at the a-SiC:H/NESA (SnO_2) interface. Here, NESA glass is a product name of Corning Inc. It was thought that by combining an extremely thin a-SiC:H film on a NESA (SnO_2) electrode and organic semiconductor films (Fig. 2.1a), the spectral sensitivity of the multiplied photocurrent generated at the NESA (SnO_2)/a-SiC:H interface could be intentionally tuned according to the various absorption regions of organic semiconductors. Thus, the holes photogenerated in the organic semiconductor films could be transported to the NESA/a-SiC:H interface and would trigger the multiplication process occurring at the NESA/a-SiC:H interface. Figure 2.1b shows the expected hole and electron movements based on the energy diagram during the discussion at that time. Figure 2.1b is a Lab notebook at that time. It was expected that the photogenerated holes in the valence band of the Me-PTC film would enter the a-SiC:H film and accumulate at the NESA/a-SiC:H interface, and the tunneling injected electrons from NESA to a-SiC:H would enter the conduction band of Me-PTC. I expected that the multiplied photocurrent injected at the NESA/a-SiC:H interface could be caused by the holes photogenerated in the Me-PTC film.

As expected, relatively large photocurrent multiplication was observed for a-SiC:H film thicknesses of 20, 30, and 50 nm (Fig. 2.1c). Figure 2.1c is a Lab notebook at that time. the cells were irradiated with 600 nm monochromatic light. As expected, the spectral sensitivity extended to the visible region of the Me-PTC film. However, a cell without a-SiC:H (Fig. 2.1a, 0 nm) showed multiplied photocurrent (Fig. 2.1c), which was not expected. Surprisingly, a-SiC:H is not necessary, and the Me-PTC film itself can cause photocurrent multiplication. The multiplication rate was higher than that in the a-SiC:H case. This was an accidental discovery, and was the first

M. Hiramoto, *Organic Photocurrent Multiplication*, Electronic Materials: Science & Technology, https://doi.org/10.1007/978-981-99-1237-7_2

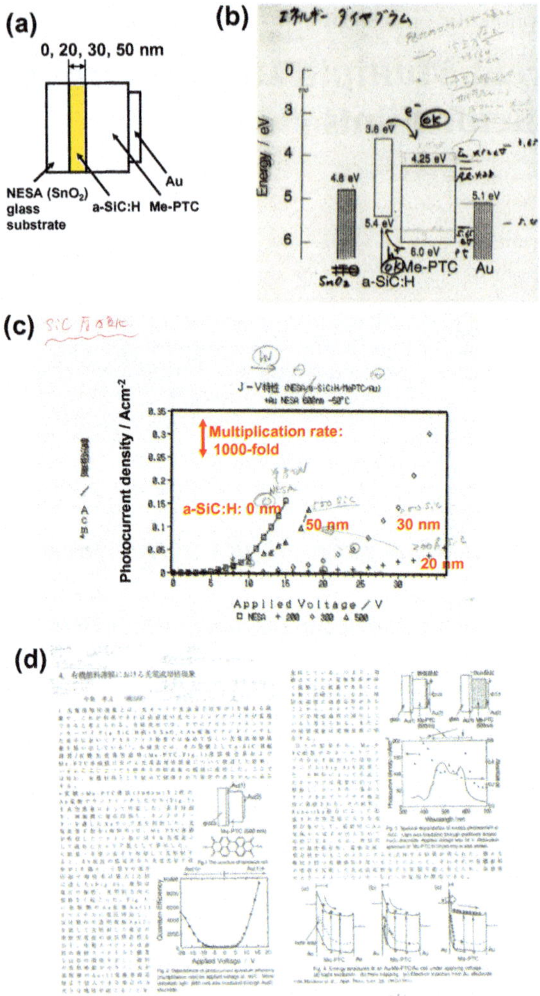

Fig. 2.1 **a** Cell structure combining a very thin a-SiC:H film and Me-PTC film. **b** Expected hole and electron movements based on the energy diagram illustrated during discussion before experiment. The conduction bands and valence bands of a-SiC:H and Me-PTC, and the work functions of metal electrodes are illustrated. This is a Lab notebook at that time. **c** Rare plots of the first observation of photocurrent multiplication. The relationship between the photocurrent density and applied voltage is shown in the raw data. The NESA (SnO_2) electrode was biased negatively with respect to the Au electrode. The thickness of a-SiC:H was decreased in the order of 50, 30, 20, and 0 nm. Even without a-SiC:H (0 nm), a multiplication rate higher than 1,000-fold was observed. Only Me-PTC was excited by 600 nm monochromatic light. This is a Lab notebook at that time. **d** Abstract (in Japanese) for the master's course presentation of Imahigashi in 1992 [4]. He was the author's student who fabricated the first organic photocurrent multiplication device

observation of photocurrent multiplication in an organic semiconductor film. The abstract of the master's thesis (in Japanese) of the student who first discovered this multiplication in 1992 is shown in Fig. 2.1d.

The idea of combining the a-SiC:H/SnO_2 interface and Me-PTC is rather straightforward. However, this idea led to an accidental discovery. At that time, the author related photocurrent multiplication in inorganic semiconductors using a-SiC:H as well as organic solar cells using Me-PTC. This was an accidental situation and guided the author to this accidental discovery. This is an example of serendipity. The author strongly recommends that young scientists should not hesitate to execute simple ideas. For this, a strong willpower is necessary, since most ordinary people with no strong motivation do not always execute such an idea. Such people always use the discouraging sentence "It does not work anyway."

2.2 Photocurrent Multiplication in *n*-Type Organic Semiconductor Films

2.2.1 Multiplication Rate Reaching Ten Thousand-Fold

Photocurrent multiplication was first observed in an *n*-type perylene pigment (N-methyl-3,4,9,10-perylenetetracarboxyl-diimide, Me-PTC, Fig. 2.2) [4]. The perylene derivative series are well-known photoconductors that have been used as charge-generating materials for electrophotography [5, 6] and organic solar cells [7–12]. The Me-PTC sample was purified by train sublimation [13].

The cell structure used in this study is shown in Fig. 2.2. A 500 nm-thick Me-PTC film was sandwiched between two semi-transparent Au electrodes. The front

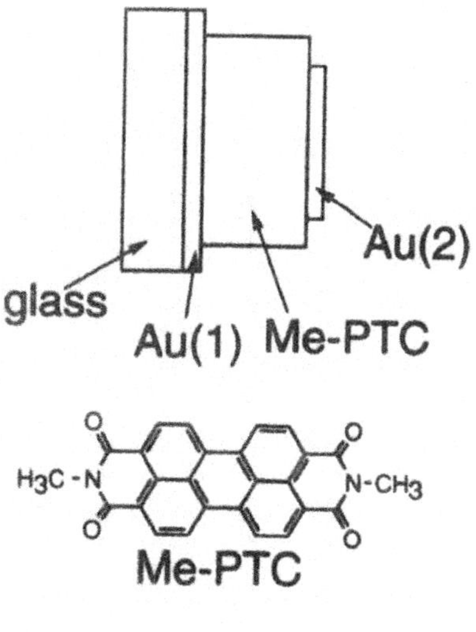

Fig. 2.2 Cell structure and chemical formula of Me-PTC. Reproduced with the permission of [4]. © 1994 American Institute of Physics

electrode [Au(1)] was deposited on a glass substrate by sputtering. The Me-PTC and back Au electrode [Au(2)] were deposited by vacuum evaporation at a pressure of 1 $\times$ 10^{-3} Pa at a deposition rate of 0.1 nm s^{-1}. The cells were set in an optical cryostat evacuated to 0.1 Pa. The photocurrent was measured using an electrometer (Keithley 485) with a voltage applied between the two Au electrodes. Monochromatic light from a 500 W Xe lamp through a monochromator was used as the light source. The light intensity was determined using a silicon photodiode (Hamamatsu Photonics, S1337-66BQ). The photocurrent quantum efficiency, that is, the multiplication rate, is defined as the ratio of the number of photogenerated carriers collected to the number of photons absorbed by the Me-PTC film.

Figure 2.3 shows the dependence of the multiplication rate on the applied voltage at −50 °C. Monochromatic light of 600 nm was irradiated on the Au(2) electrode. The voltage axis corresponds to that of the Au(1) electrode. When Au(2) was negatively biased, the multiplication rate exceeded unity at 1.4 V and increased sharply above 5 V, reaching approximately 10,000 at 16 V. Please note that this value is not expressed by a percentage; it implies 10,000-fold, i.e., ten thousand-fold, in multiplication rate. The photocurrent and dark current densities at 16 V were 720 mAcm^{-2} and 50 mAcm^{-2}, respectively. A high multiplication rate of 5,000-fold was observed when the opposite Au(1) electrode was negatively biased to −20 V. This device could be cycled on and off at least 10 times without degradation, even under a high current density of the order of Acm^{-2}. Very high photocurrent multiplication was observed at room temperature. By cooling the device, the multiplication rate increased and reached a maximum at −50 °C. Further cooling reduced the multiplication rate. Therefore, almost all measurements were performed at −50 °C.

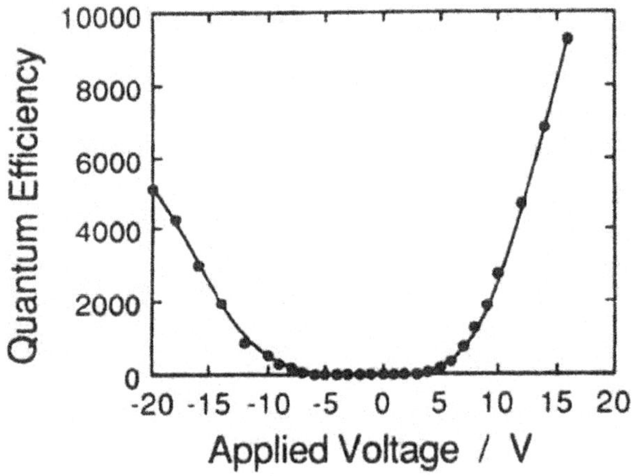

Fig. 2.3 Dependence of photocurrent quantum efficiency on applied voltage (V_{appl}). V_{appl} was applied to the Au(1) electrode. Monochromatic light of wavelength 600 nm was irradiated from the Au(2) side. The light intensity through the Au electrode was 0.16 mWcm^{-2}. Measurements were performed at −50 °C. Reproduced with the permission of [4]. © 1994 American Institute of Physics

2.2.2 Identifying the Interface

The spectral dependence of the multiplied photocurrent density and absorption spectrum of the Me-PTC film are shown in Fig. 2.4. The Au(1) electrode was negatively biased. When light was irradiated through the Au(2) electrode, the photocurrent action spectrum (Fig. 2.4b, curve A) showed sharp peaks at the edges and dips of two absorption peaks for Me-PTC absorption (Fig. 2.4a). Because the film transmission increases sharply on each side of the absorption, the delivery of incident photons through the absorbing layer to the negatively biased Au(1)/Me-PTC interface region is maximized. This is the masking effect. In contrast, when light was irradiated through the negatively biased Au(1) electrode, the photocurrent action spectrum showed few peaks (Fig. 2.4b, curve B), although it still showed a weak reverse relationship. These results suggest that the observed multiplication occurred near the negatively biased Au electrode. This interpretation was further supported by the fact that completely reverse behavior was observed for the photocurrent action spectrum when the opposite Au(2) electrode was negatively biased. Sharp peaks, such as curve A, were observed by irradiating the Au(1) electrode, and small peaks, such as curve B, were observed by irradiating the Au(2) electrode.

2.2.3 Transient Response

The multiplied photocurrent always showed a transient response of the order of several tens of seconds for light on and off (Fig. 2.5). This suggests that the multiplication process is related to a carrier trapping event.

2.2.4 Multiplication Mechanism—Tunneling Injection of Electrons

A multiplication mechanism is proposed here assuming that there are neutral hole traps within the band gap of the Me-PTC film. Figure 2.6 illustrates the energy structure of the cell under biasing. The applied electric field is mainly distributed in the highly resistive depletion layer formed between the negatively biased Au and Me-PTC, because there is a Schottky barrier at the Au/Me-PTC interface, and the Me-PTC film behaves as an n-type semiconductor [14, 15]. Under light irradiation, the photogenerated electrons and holes move along the potential gradient (Fig. 2.6a). The holes are then captured by traps (Fig. 2.6b). The accumulation of trapped holes leads to the development of a strong electric field across the depletion layer. Finally, the tunneling injection of electrons occurs from the Au electrode through the shrunken depletion layer (Fig. 2.6c).

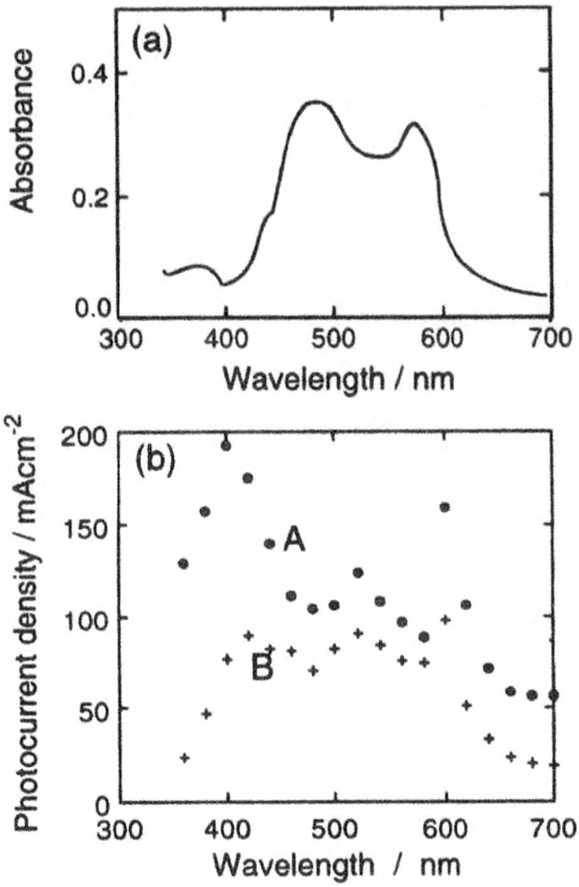

Fig. 2.4 **a** Absorption spectrum of Me-PTC film (40 nm). **b** The spectral dependence of the multiplied photocurrent density at −50 °C. Au(1) was biased with −10 V with respect to Au(2). Au(2) (curve A) and Au(1) (curve B) were irradiated with monochromatic light. Light intensities passing through the Au(1) and Au(2) electrodes were adjusted to the same value (0.16 mWcm^{-2} at 600 nm). Reproduced with the permission of [4]. © 1994 American Institute of Physics

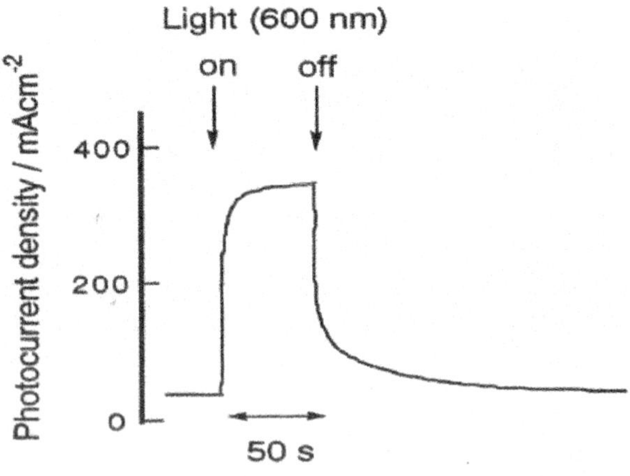

Fig. 2.5 Typical response of the multiplied photocurrent. The multiplication rate was approximately 1,600-fold

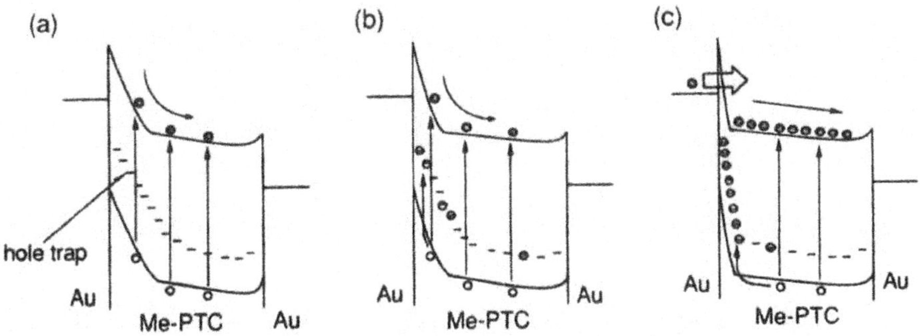

Fig. 2.6 Energy structures of an Au/Me-PTC/Au cell under an applied electric field. **a** Light irradiation. **b** Hole trapping. **c** Electron injection through the shrunken depletion layer from the negative Au electrode. Reproduced with the permission of [4]. © 1994 American Institute of Physics

In this model, the transient response of the multiplied photocurrent is explained by the hole capturing and emission processes. The multiplication enhancement by cooling to −50 °C is due to the increase in the number of trapped holes. The reduced multiplication by further cooling was caused by the increase in the resistivity of the Me-PTC bulk, which prevented the concentration of the electric field at the Au/Me-PTC interface.

2.2.5 Hole Traps—Origin of Multiplication

The existence of a hole trap is the origin of the surprisingly large photocurrent multiplication, reaching 10^4-fold. However, at this stage, the occurrence of hole traps, as illustrated in Fig. 2.6, had merely been an assumption. Initially, the large number of unknown impurities were believed to be acting as hole traps in organic pigment films owing to the difficulty of purification. However, a detailed study revealed that hole traps are not impurities (Chaps. 3 and 4).

2.3 Photocurrent Multiplication in *p*-Type Organic Semiconductor Films

Large photocurrent multiplication due to hole injection reaching 2,500-fold was observed in organic p-type quinacridone pigment films [16]. This phenomenon was revealed to be due to photoinduced hole injection from a positively biased metal electrode to the p-type organic film through its depletion layer under a high electric field, which was built up by the photoaccumulation of trapped electrons near the interface. This is a clear example of the reverse mechanism of photoinduced electron

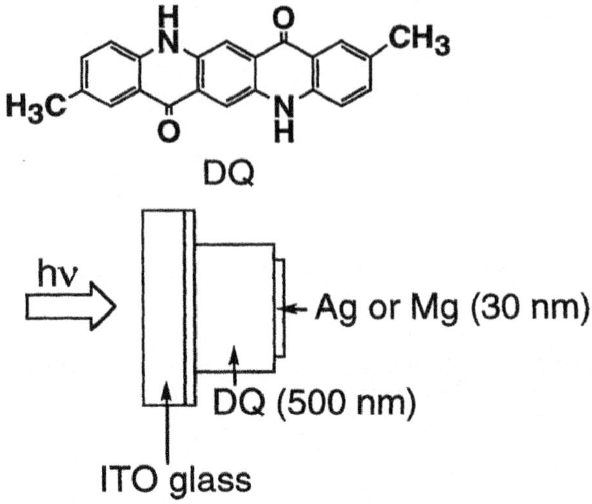

Fig. 2.7 Cell structure and chemical formula of DQ. Reproduced with the permission of [16]. © 1996 American Institute of Physics

injection observed in *n*-type organic pigment films, which can occur in *p*-type organic pigment films.

2.3.1 Multiplication Behavior

The quinacridone pigment (2,9-dimethyl quinacridone, DQ, Fig. 2.7) was purified twice by train sublimation [13]. The cell structure used in this study is shown in Fig. 2.7. A 500 nm-thick DQ film was sandwiched between indium tin oxide (ITO) and metal electrodes. Remarkably large photocurrent multiplication was observed in *p*-type DQ pigment films. Figure 2.8 shows the dependence of the photocurrent quantum efficiency on the applied voltage at room temperature. The voltage axis refers to the Ag or Mg electrodes. When Ag was positively biased, the quantum efficiency exceeded 1 V and reached approximately 2,500-fold at 20 V at room temperature (curve A). The photocurrent and dark current densities at 20 V were 174 $mAcm^{-2}$ and 48 $mAcm^{-2}$, respectively. When the ITO electrode was positively biased, a multiplication rate of 1500-fold was observed. The multiplication rate was decreased by cooling the device.

2.3.2 Influence of the Kinds of Electrode Metals

The multiplication behavior was significantly affected by the type of metal used for the positively biased electrode. When the Ag electrode was replaced with a Mg electrode, the photocurrent quantum efficiency decreased considerably, and a higher

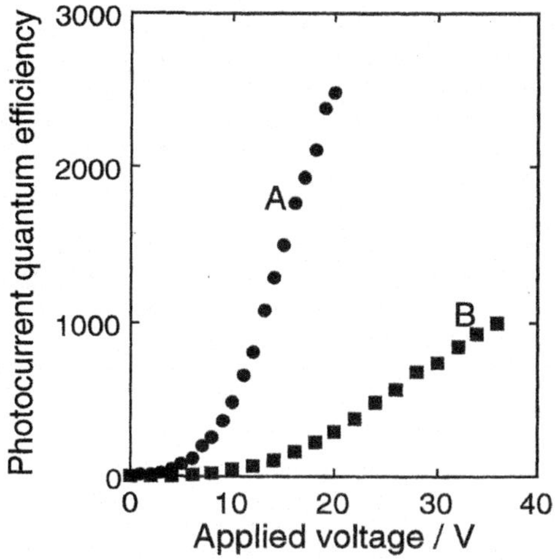

Fig. 2.8 Dependence of photocurrent quantum efficiency on applied voltage. Ag (curve A) or Mg (curve B) was used as the positively biased metal electrode. Voltage was applied to the Ag or Mg electrode. The intensity of monochromatic light (600 nm) irradiating the ITO electrode was 0.166 $mW.cm^{-2}$. Measurements were performed at room temperature. Reproduced with the permission of [16]. © 1996 American Institute of Physics

voltage of 36 V was required to obtain a multiplication rate of 1,000-fold (Fig. 2.8, curve B). In contrast, when the electrodes were negatively biased, such a replacement had little effect on the multiplication behavior. These results strongly suggest that the multiplication phenomenon in p-type DQ films also occurs at the interface with the metal electrode, similar to n-type organic pigments [4]. Note, however, that the active interface responsible for the multiplication is in marked contrast to the case of the n-type pigments, where the multiplication behavior was affected only by the type of the negatively biased electrode [4].

2.3.3 *Specifying the Interface*

The active interface can be identified from the photocurrent action spectrum. Figure 2.9 shows the spectral dependence of the multiplied photocurrent (dots) and the absorption spectrum of the DQ film (solid curve). The Ag electrode was positively biased and the ITO electrode was irradiated with monochromatic light. The multiplied photocurrent showed maxima in the wavelength region of the absorption edges of the DQ film, where the incident photons deeply penetrated the DQ film and approached the opposite interface of the positively biased Ag electrode. It also showed a minimum in the wavelength region of strong absorption, where most of the incident photons were absorbed near the interface of the negatively biased ITO electrode. However, when the positively biased ITO electrode was illuminated, such distinct peaks were hardly observed. These results led us to conclude that the interface with the positively biased electrode is responsible for the multiplication process in the p-type DQ pigment.

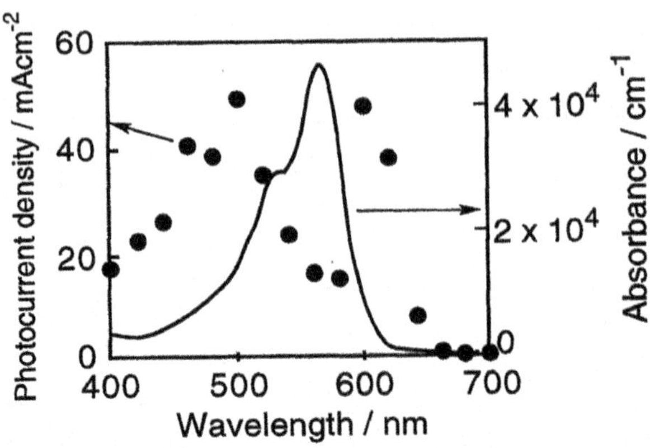

Fig. 2.9 Spectral dependence of excess photocurrent (dots) for a cell with the ITO/DQ(650 nm)/Ag(30 nm) structure. Ag was biased at +15 V with respect to ITO. The intensity of the light illuminating the ITO glass substrate was 0.18 mWcm^{-2} at a wavelength of 600 nm. The absorption spectrum of the 500-nm-thick DQ film (solid curve) is also shown. Reproduced with the permission of [16]. © 1996 American Institute of Physics

2.3.4 *Transient Response*

The multiplied photocurrent exhibited a rather slow transient response of the order of 10–20 s to the turning on and off of light. This suggests that the multiplication is also closely related to some of the trapping events of the photogenerated carriers.

2.3.5 *Multiplication Mechanism—Tunneling Injection of Holes*

Accordingly, we propose the following multiplication mechanism, assuming the presence of a neutral electron trap within the DQ. Figure 2.10 illustrates the energy structures of the cell under an applied bias positive to the Ag electrode. Since the formation of a Schottky barrier between the *p*-type DQ and Ag electrode has been confirmed [17, 18], the applied field is mainly distributed in such a highly resistive depletion layer. Under light irradiation, the photogenerated electrons move along the potential gradient and are trapped within the bandgap (a). A high electric field across the depletion layer, which is built up by the accumulation of electrons trapped near the positively biased electrode, causes the tunneling injection of holes from the Ag electrode through the shrunken depletion layer (b). This photoinduced hole injection occurring at the interface between the *p*-type DQ and a positively biased electrode is the reverse mechanism to the photoinduced electron injection occurring at the interface between the *n*-type perylene pigment film and a negatively biased electrode, as reported previously [4].

Fig. 2.10 Energy structures of an ITO/DQ/Ag cell under a positive electric field applied to Ag. **a** Trapping of photogenerated electrons. **b** Hole injection from the positively biased Ag electrode. Reproduced with the permission of [16]. © 1996 American Institute of Physics

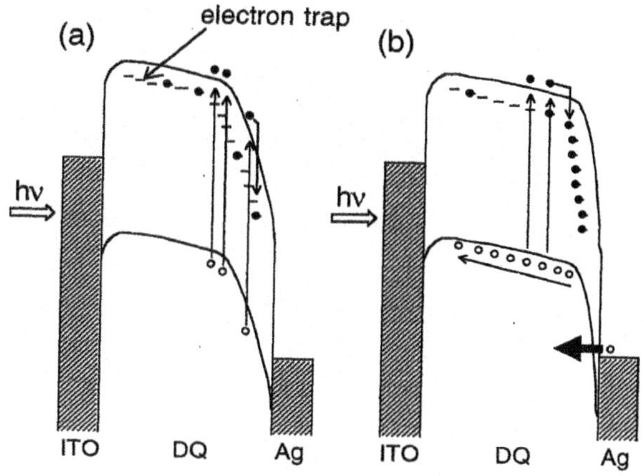

Multiplication was suppressed by replacing the metal of a large work function (Ag, 4.26 eV) [19] with a metal of a smaller work function (Mg, 3.66 eV), as shown in Fig. 2.8. Hole injection in the *p*-type film must be suppressed because of the larger energy barrier formed by a metal with a smaller work function. This result is in clear contrast to that observed for *n*-type perylene pigment films, in which multiplication was decreased by replacing a small-work-function metal (Au, 5.1 eV) with a larger-work-function metal (Pt, 5.65 eV) [19]. These findings are consistent with both the mechanisms of electron and hole injection proposed for *n*-type and *p*-type films, respectively. The interface at which the high electric field was concentrated was found to be opposite for the *p*- and *n*-type films. This seems to be due to the opposite direction of band bending at the metal interface, which decreases the tunneling width of injection for only a single type of charge carrier.

2.4 Generality

In Fig. 2.11, organic semiconductors exhibiting photocurrent multiplication are presented. Photocurrent multiplication was observed in most of the organic films examined, irrespective of the *n*- and *p*-type organic semiconductors used. Thus, photocurrent multiplication is a general phenomenon in organic semiconductors. The observed multiplication rate of 10^5-fold was surprisingly high. It should be noted that this value is not a percentage. This value is an integer.

In the case of *n*-type organic semiconductors, perylene pigments, fullerene (C_{60}), and naphthalene derivatives (naphthalene tetracarboxylic anhydride, NTCDA) showed large photocurrent multiplication. In perylene pigments (Me-PTC, n-Bu-PTC, PTCDA, PhEt-PTC, and t-BuPh-PTC), the multiplication behavior was dependent on the type of substituent. Photocurrent multiplication was observed even at room temperature. A typical example is a naphthalene derivative, which showed

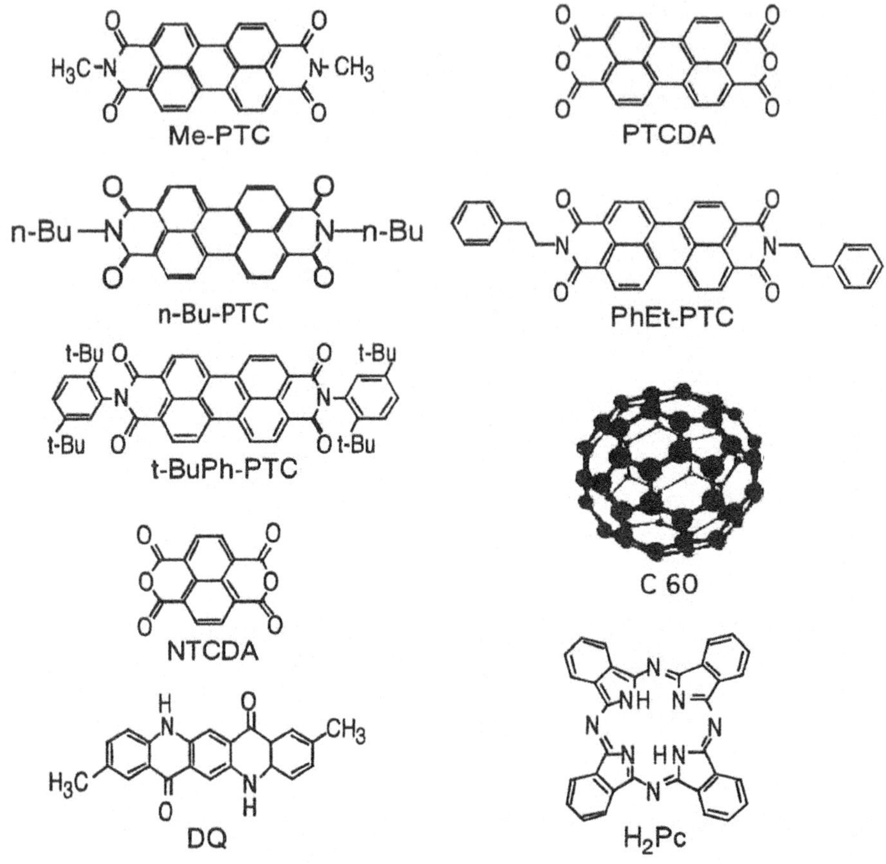

Fig. 2.11 Organic semiconductors showing photocurrent multiplication

the largest photocurrent multiplication exceeding 10^5-fold at room temperature [20] (Fig. 2.12). This is contrary to the case of Me-PTC, in which the maximum multiplication rate was observed under cooling at −50 °C.

A multiplication rate exceeding 10^3-fold was observed in *p*-type phthalocyanine (H_2Pc, TiOPc, and CuPc) and quinacridone (DQ) pigments. For *p*-type cases, all multiplications are caused by the mechanism reverse to that of *n*-type perylene and NTCDA films, that is, photoinduced hole injection from a positively biased metal electrode (Fig. 2.10).

Because various organic semiconductors with different absorption spectra show photocurrent multiplication, multiplication-type color-sensing devices with sensitivity to various wavelength regions can be fabricated. NTCDA cells responded selectively to blue light between 380 and 480 nm, whereas Me-PTC cells responded to the entire visible region. CuPc cells can respond to the near-infrared region.

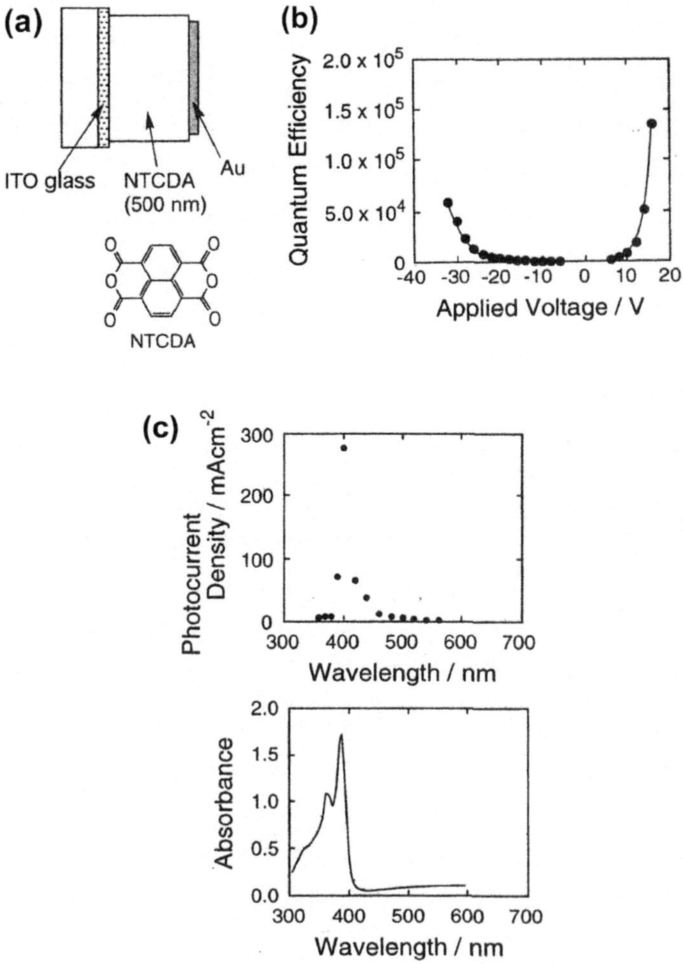

Fig. 2.12 **a** Cell structure and chemical formula of NTCDA. **b** Dependence of photocurrent quantum efficiency on applied voltage for a cell shown in Fig. 2.12a. For a positive applied voltage, the structure is denoted as ITO (+)/NTCDA/Au (−). For a negative applied voltage, the structure was denoted as ITO (−)/NTCDA/Au (+). The Au electrode was irradiated with monochromatic light of wavelength 400 nm (40 μWcm^{-2}). Measurements were performed at room temperature. **c** (Upper) Spectral dependence of multiplied photocurrent density. The ITO electrode was irradiated with monochromatic light. The Au electrode was negatively biased at 16 V. (Lower) Absorption spectra of NTCDA film (100 nm). Reproduced with the permission of [20]. © 1996 American Institute of Physics

2.5 Room-Temperature Multiplication

At room temperature, the largest photocurrent multiplication of 130,000-fold was observed in NTCDA films [20, 21]. In this section, the multiplication behavior of NTCDA is summarized.

2.5.1 *Multiplication Rate*

Sandwich-type cells, in which the NTCDA was sandwiched between Au and ITO (Fig. 2.12a), were fabricated. Figure 2.12b shows the dependence of the multiplication rate on the applied voltage. When the Au electrode was negatively biased, the photocurrent quantum efficiency (i.e., the multiplication rate) exceeded unity below 1 V and reached 134,600-fold at 16 V. This multiplication is an integer number and not a percentage. The photocurrent and dark current densities at 16 V were 460 and 14 $mAcm^{-2}$, respectively. A high multiplication rate exceeding 59,000-fold was also observed when the ITO electrode was negatively biased to −32 V. These multiplication rates are comparable to those of a photomultiplier tube. It should be noted that the observed multiplication rates were measured at room temperature. The multiplication rate in this device showed maxima at approximately 30 and −80 °C. Thus, cooling was not necessary for photocurrent multiplication. The responses of the multiplied photocurrent were quite reproducible after many on/off cycles of light and voltage, despite the high current density of several hundred $mAcm^{-2}$. However, when the organic layer thickness was decreased from 500 to 300 nm, the contrast between the photo and dark regions decreased owing to the increase in the dark current, although the responses were still reproducible.

2.5.2 *Spectral Sensitivity*

Figure 2.12c shows the spectral dependence of the multiplied photocurrent density. The positively biased ITO electrode was irradiated with light. This cell responded selectively to UV-blue light between 380 and 460 nm, and showed a sharp peak at 400 nm, corresponding to the absorption tail of the NTCDA film with a peak at 385 nm (Fig. 2.12c, lower figure). This is due to the masking effect, which was also observed for the Me-PTC film (Sect. 2.2.2). Because the incident light in the region of the absorption tail can penetrate deeper into the NTCDA film compared to that at the absorption peak, carriers are photogenerated nearer to the Au electrode and are more easily delivered to the negatively biased interface. This result indicates that multiplication occurs at the negatively biased interface. Because the transparent NTCDA can act as a window for visible to infrared light, the spectral sensitivity can be intentionally tuned by combining NTCDA with various kinds of organic semiconductors [20, 21].

2.5.3 *Influence of the Kinds of Metals*

The type of metal used for the negatively biased electrode strongly influences the multiplication behavior. By replacing the Au electrode with a Ag electrode of a smaller work function, the multiplication rate can be increased to 2.5 times over the entire voltage region. This result again indicates that multiplication occurs near the interface between the NTCDA film and the negatively biased metal electrode.

2.5.4 *Multiplication Mechanism*

The multiplication mechanism is explained based on the same mechanism as that for Me-PTC (Fig. 2.6). NTCDA showed *n*-type behavior, and Schottky barriers were observed at the interfaces of NTCDA/Au and ITO/NTCDA without applying any bias under dark conditions (Fig. 2.6a). For a single NTCDA cell, when UV-blue light is applied, the photogenerated carriers move along the potential gradient, and some of the holes are captured by the hole traps (Fig. 2.6b). The accumulation of trapped holes near the NTCDA/Au interface triggers the concentration of the applied electric field at the negatively biased interface. Consequently, numerous electrons are injected from the Au electrode via tunneling.

We excluded the mechanism based on Zener tunneling [22] from the valence band because of the strong dependence of the multiplication behavior on the type of metal. In addition, we excluded the mechanism based on avalanche breakdown [23, 24] because of the low electron drift mobility [25] of the Me-PTC pigment (less than 10^{-3} cm^2V^{-1} s^{-1}), which can hardly attain enough kinetic energy to cause impact ionization.

2.6 Conclusion

Photocurrent multiplication was first observed in vacuum-deposited organic semiconductor films sandwiched between two metal electrodes at room temperature. The multiplication rate was more than 10^6-fold, that is, over 100,000 electrons flowed through the device by one irradiated photon on average. For *n*-type organic semiconductors, multiplication occurs by the tunneling injection of electrons from the negatively biased metal electrode to the organic film caused by the electric field concentration at the organic/metal interface due to the accumulation of trapped holes near the organic/metal interface. For *p*-type organic semiconductors, the opposite mechanism occurs, namely, multiplication occurs by the tunneling injection of holes from the positively biased metal electrode to the organic film caused by the electric field concentration at the organic/metal interface due to the accumulation of trapped electrons near the organic/metal interface. Multiplication phenomena by the same tunneling mechanism have been observed for many organic semiconductors and can be regarded as a general phenomenon in organic semiconductor films. This novel phenomenon can be utilized in a multiplication-type organic photodetector.

References

1. Hiramoto, M., Yoshimura, K., Nakayama, Y., Akita, S., Kawamura, T., Yokoyama, M.: Photocurrent multiplication in amorphous silicon carbide films. Appl. Phys. Lett. **59**, 1992–1994 (1991)
2. Hiramoto, M., Yoshimura, K., Yokoyama, M.: Photomodulation of photocurrent multiplication in a high gain amorphous silicon carbide film. Appl. Phys. Lett. **59**, 1992–1994 (1991)
3. Hiramoto, M., Yoshimura, K., Yokoyama, M.: Photocurrent multiplication in amorphous silicon carbide films. J. Imag. Sci. Technol. **37**, 192–196 (1993)
4. Hiramoto, M., Imahigashi, T., Yokoyama, M.: Photocurrent multiplication in organic pigment films. Appl. Phys. Lett. **64**, 187–189 (1994)
5. Loutfy, R.O., Hor, A.M., Kazmaler, P., Tam, M.: Layered organic photoconductive (OPC) devices incorporating N, N'-disubstituted diimide and bisarylimidazole derivatives of perylene-3,4,9,10-tetracarboxylic acid. J. Imaging Sci. **33**, 151–159 (1989)
6. Nakazawa, T., Muto, N., Mizuta, Y., Kawahara, A., Miyamoto, E., Tsutsumi, M., Ikeda, T., Nagahashi, K.: A new type monolayered organic photoconductor for positive charging xerography—chemistry and characteristics. Nihon Kagaku Kaishi. **1992**, 1007–1018 (1992)
7. Tang, C.W.: Two-layer organic photovoltaic cell. Appl. Phys. Lett. **58**, 1062–1064 (1986)
8. Hiramoto, M., Kishigami, Y., Yokoyama, M.: Doping effect on the two-layer organic solar cell. Chem. Lett. **1990**, 119–122 (1990)
9. Hiramoto, M., Suezaki, M., Yokoyama, M.: Effect of thin gold interstitial-layer on the photovoltaic properties of tandem organic solar cell. Chem. Lett. **1990**, 327–330 (1990)
10. Hiramoto, M., Fujiwara, H., Yokoyama, M.: Three-layered organic solar cell with a photoactive interlayer of codeposited pigments. Appl. Phys. Lett. **58**, 1062–1064 (1991)
11. Hiramoto, M., Fujiwara, H., Yokoyama, M.: p-i-n like behavior in three-layered organic solar cells having a co-deposited interlayer of pigments. J. Appl. Phys. **72**, 3781–3787 (1992)
12. Hiramoto, M., Fukusumi, H., Yokoyama, M.: Organic solar cell based on multistep charge separation system. Appl. Phys. Lett. **61**, 2580–2582 (1992)
13. Wagner, H.J., Loutfy, R.O., Hsiao, C.: Purification and characterization of phthalocyanines. J. Mater. Sci. **17**, 2781–2791 (1982)
14. Hiramoto, M., Ihara, K., Yokoyama, M.: Fermi level shift in photoconductive organic pigment films measured by Kelvin vibrating capacitor method. Jpn. J. Appl. Phys. **34**, 3803–3807 (1995)
15. Hiramoto, M., Ihara, K., Fukusumi, H., Yokoyama, M.: Conduction type control from n- to p-type for organic pigment film purified by reactive sublimation. J. Appl. Phys. **78**, 7153–7157 (1995)
16. Hiramoto, M., Kawase, S., Yokoyama, M.: Photoinduced hole injection multiplication in p-type quinacridone pigment films. Jpn. J. Appl. Phys. **35**, L349–L351 (1996)
17. Hiramoto, M., Nakamura, H., Shiojiri, T., Yokoyama, M.: Doping effects of organic solar cells using quinacridone pigment. Kobunshi Ronbunshu **47**, 915–920 (1990). [in Japanese]
18. Manabe, K., Kusabayashi, S., Yokoyama, M.: Long-life organic solar cell fabrication quinacridone pigments. Chem. Lett. **1987**, 609–612 (1987)

19. Michaelson, H.B.: The work function of the elements and its periodicity. J. Appl. Phys. **48**, 4729–4733 (1977)
20. Katsume, T., Hiramoto, M., Yokoyama, M.: Photocurrent multiplication in naphthalene tetracarboxylic anhydride film at room temperature. Appl. Phys. Lett. **69**, 3722–3724 (1996)
21. Hiramoto, M., Yokoyama, M.: Organic photocurrent multiplication devices. Jpn. Patent 100797 (2002)
22. Bardeen, J.: Phys. Rev. Lett. **6**, 57 (1961)
23. Sze, S.M.: Physics of Semiconductor Devices. Chap. 5, p. 112, Wiley (1969)
24. Sze, S.M., Gibbons, G.: Avalanche breakdown voltages of abrupt and linearly graded p-n junctions in Ge, Si, GaAs, and GaP. Appl. Phys. Lett. **8**, 111 (1966)
25. Magin, E.H., Borsenberger, P.M.: Electron transport in N, N'-bis(2-phenethyl)-perylene-3,4: 9,10-bis(dicarboximide). J. Appl. Phys. **73**, 787–791 (1993)

Chapter 3
Analyses of Multiplication Behaviors—Structural Trap

3.1 Multiplication Mechanism

We observed large photocurrent multiplication in various organic films, including a series of perylene pigments [1], a quinacridone pigment [2], and a naphthalene derivative [3] (Chap. 2). In particular, Me-PTC, one of the perylene pigments, showed very large photocurrent multiplication by a factor of more than 10^4 (Fig. 3.1b). To explain the photocurrent multiplication, a mechanism based on the tunneling injection of electrons from a metal electrode to the pigment layer in a high electric field resulting from the photoaccumulation of space charges of trapped holes near the interface was proposed (Fig. 3.3) [1–3].

In this chapter, detailed analyses of multiplication behavior based on transient photocurrent measurements, Fowler–Nordheim (FN) plots, and thermally stimulated current (TSC) measurements are summarized. A structural trap model is proposed to suggest a trap candidate.

3.2 Transient Response

3.2.1 Two Components of Photocurrent

The proposed mechanism of photocurrent multiplication can be regarded as successive hole accumulation and electron tunneling injection steps. This strongly suggests that observing the initial stage of transient photocurrent is important. Therefore, transient photocurrent measurements were performed [4, 5]. Figure 3.1a shows a block diagram of the transient photocurrent measurements. A positively biased indium tin oxide (ITO) electrode was irradiated with an orange light-emitting diode (LED) emitting at 620 nm (FWHM: 20 nm). The LED was driven by the step voltage from a function generator. The transient photocurrent response was measured using a high-speed current amplifier (Keithley 428) connected to a digital oscilloscope (Nicolet

M. Hiramoto, *Organic Photocurrent Multiplication*, Electronic Materials: Science & Technology, https://doi.org/10.1007/978-981-99-1237-7_3

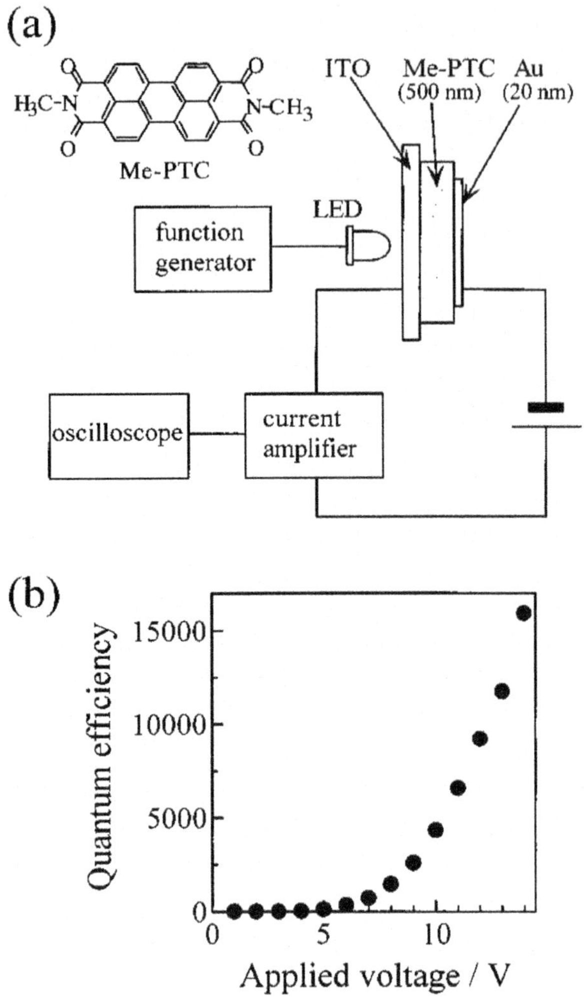

Fig. 3.1 **a** Block diagram for the transient photocurrent measurements. The structure of the photocurrent multiplication device and the chemical formula of the perylene pigment (Me-PTC) are also shown. The cell was set in an optical cryostat evacuated to 10^{-3} Torr. Measurements were carried out at 203 K. **b** Dependence of the multiplication rate on the applied voltage for the cell in Fig. 3.1a. The ITO electrode was irradiated with continuous light (600 nm, 40 μWcm^{-2}). The Au electrode was biased negatively with respect to the ITO electrode. Reproduced with the permission of [4]. © 1998 American Institute of Physics

3091). The present cell (Fig. 3.1a) showed a multiplication rate of 1.5×10^4 when Au was negatively biased at 14 V with respect to the ITO electrode (Fig. 3.1b).

Figure 3.2 shows the initial transient response of the photocurrent observed at −70 °C for three different light intensities [4, 5]. The inset shows a typical response to saturation. The transient photocurrent consisted of two components. The first component rose within 10 μs and reached a plateau. The second component rose drastically after an onset time of 43 ms at a light intensity of 0.19 $mWcm^{-2}$ (Fig. 3.2, curve c, vertical arrow).

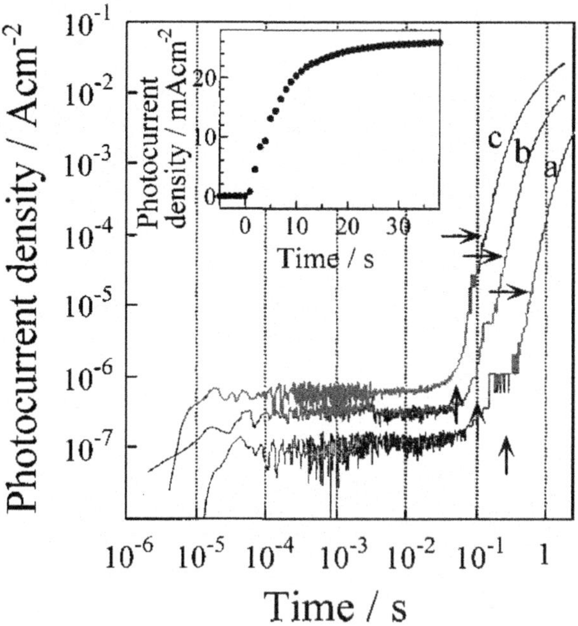

Fig. 3.2 Initial stages of the transient photocurrent curves for light intensities of **a** 36, **b** 95, and **c** 190 μWcm^{-2}. Both axes have a logarithmic scale. Measurements were performed at 203 K. The Au electrode was negatively biased by 15 V with respect to the ITO electrode. The vertical and horizontal arrows in the figure indicate the onset time and the point at which the photocurrent quantum efficiency equals unity, respectively. Inset is the linear scale plot of **b**. Reproduced with the permission of [4]. © 1998 American Institute of Physics

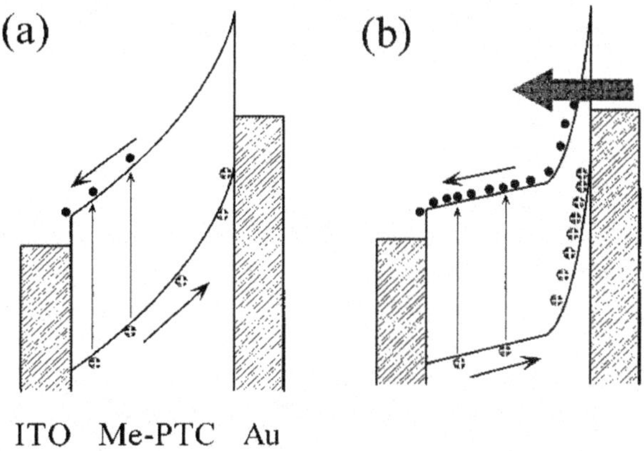

Fig. 3.3 Energy structures of the photocurrent multiplication device under light irradiation. The Au electrode was negatively biased with respect to the ITO electrode. **a** Before the onset time. **b** After the onset time. Reproduced with the permission of [4]. © 1998 American Institute of Physics

3.2.2 First Component: Primary Photocurrent

The photocurrent quantum efficiency of the first component in the plateau region was 0.65%. The first component of the photocurrent observed before the onset time was constant for a given voltage and light intensity and showed a quantum efficiency far smaller than unity. Therefore, we concluded that the first component can be attributed to the primary photocurrent due to carrier photogeneration in the Me-PTC film.

3.2.3 Second Component: Tunneling Injection

The photocurrent quantum efficiency of the second component increased steeply after the onset time, exceeded unity at 130 ms (Fig. 3.2, curve c, horizontal arrow), and showed a 180-fold increase at 1 s. Moreover, the photocurrent density of the second component steeply increased with applied voltage, obeying the FN relationship owing to the tunneling injection, as discussed in Sect. 3.3. The second component can be attributed to photocurrent multiplication arising from the tunneling injection of electrons.

3.2.4 Direct Tracing of Multiplication Process

The observed transient response clearly traces the multiplication process, including the accumulation of trapped holes near the organic/metal interface during the primary photocurrent flow before onset (Fig. 3.3a) as well as the subsequent tunneling injection of electrons under the concentrated electric field resulting from hole accumulation (Fig. 3.3b). The onset time can be considered the beginning of the tunneling injection of electrons, i.e., the beginning of multiplication. Thus, the observed transient photocurrent behavior can be regarded as clear evidence for the proposed photoinduced electron tunneling mechanism for photocurrent multiplication.

3.2.5 Required Charges for Multiplication

The onset time decreased with increasing light intensity (Fig. 3.2, curves a–c). When the light intensity increased from 0.036 to 0.19 $mWcm^{-2}$, the onset time decreased from 250 to 43 ms (Fig. 3.4a). Because the light intensity and onset time are inversely proportional to each other, we calculated the total amount of charge flowing through the cell before the onset time (onset charge) by integrating the primary photocurrent until the onset time. Figure 3.4b shows the light-intensity dependence of the onset charge. The onset charge was found to be independent of the light intensity and was estimated to be 30 $nCcm^{-2}$. This implies that the number of holes that need to be supplied to the organic/metal interface for inducing electron tunneling seems to have a specific value for the present Me-PTC/Au system. Evidently, the onset charge is closely related to the number of trapped holes at the Me-PTC/Au interface.

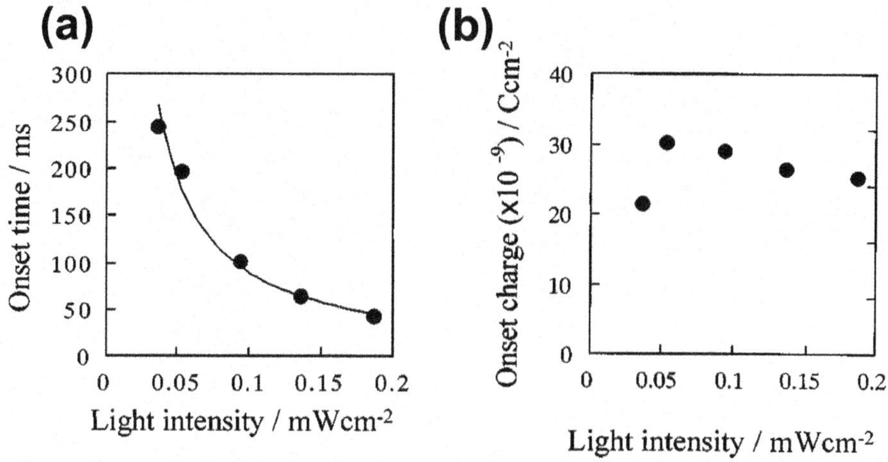

Fig. 3.4 **a** Dependence of the onset time of the second photocurrent on light intensity. The measurement conditions were the same as those described in Fig. 3.2. **b** Dependence of the amount of onset charge on light intensity. The onset charge is calculated by multiplying the onset time with the primary photocurrent density. Reproduced with the permission of [4]. © 1998 American Institute of Physics

3.3 Fowler-Nordheim (FN) Analyses

3.3.1 FN Plots of Multiplied Photocurrent [6, 7]

When the tunneling injection of electrons from the metal electrode to the conduction band of Me-PTC over a potential barrier with an energetic height of Φ_B under an applied electric field of E dominates the multiplied photocurrent (Fig. 3.5), the relationship between the observed photocurrent (J) and electric field (E) can be expressed by the following Fowler-Nordheim (FN) equation (Eq. 3.1) [8–11].

$$J = \frac{e^3}{8\pi h\phi_B} E^2 \exp\left(-\frac{8\pi\sqrt{2m*}\phi_B^{3/2}}{3eh}\frac{1}{E} \right) \tag{3.1}$$

where Φ_B and m* are the barrier height for electron injection and the effective mass of electrons in the organic layer, respectively. Based on Eq. 3.1, the relationship between $\ln(JE^{-2})$ and E^{-1} is expected to be linear.

Figure 3.6 shows the FN plot obtained from the multiplied photocurrent density (J) – voltage (V_{appl}) characteristics (inset) at −50 °C [6, 7]. A cell with an ITO/Me-PTC (500 nm)/Au structure was used (Fig. 3.1). The Au electrode was negatively biased with respect to the ITO electrode. The FN relationship was observed in the high-electric-field region during large multiplication; i.e., the $\ln(JE^{-2})$ versus E^{-1} plot fitted well with a straight line. This indicates that the tunneling process dominated the multiplied photocurrent. Based on the slope, a Φ_B value of 0.07 eV was calculated.

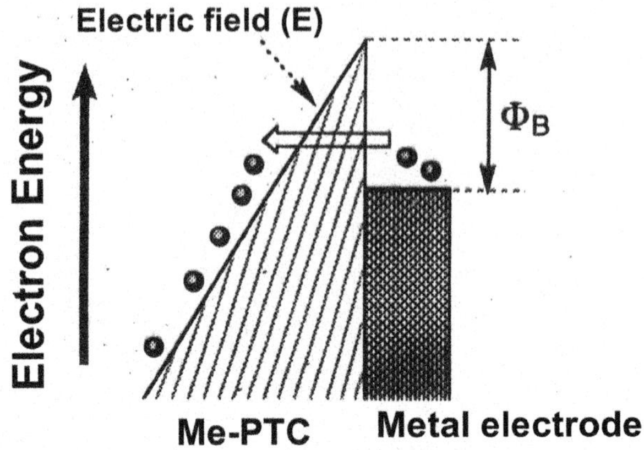

Fig. 3.5 Energetic structure of the Me-PTC/Metal electrode interface under the electric field (E). Tunneling injection of electrons from the metal electrode to the conduction band of Me-PTC over the potential barrier (Φ_B) occurs

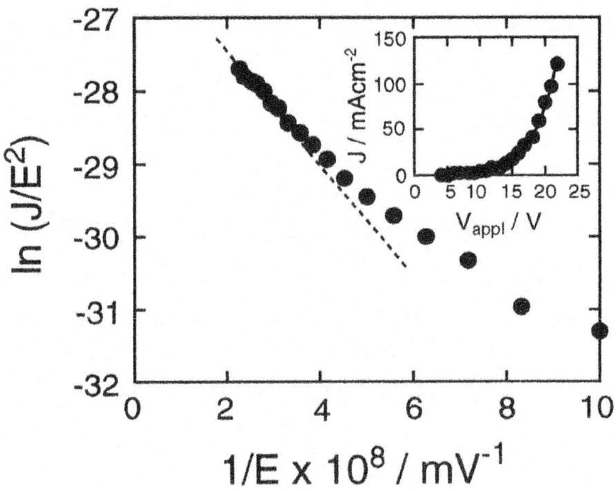

Fig. 3.6 FN plot obtained from the multiplied photocurrent (J) – voltage (V_{appl}) characteristics (inset). The applied electric field is assumed to be distributed uniformly throughout the 500-nm-thick Me-PTC film. The cell structure was ITO/Me-PTC (500 nm)/Au. The Au electrode was negatively biased with respect to ITO electrode

This value is smaller than the Φ_B value of 1.50 eV estimated independently from the work function of the Au film (4.90 eV) and from the upper edge of the valence band of the Me-PTC film (5.40 eV), which was obtained by atmospheric photoelectron emission analysis [12] and by using the optical band gap of the Me-PTC film (2.0 eV). This large discrepancy can be attributed to the improper assumption that the applied electric field is uniformly distributed throughout the Me-PTC film (Fig. 3.7a).

3.3.2 Estimation of Field Concentration Width (d)

In order to resolve this discrepancy, we assumed, based on the multiplication mechanism, that the applied electric field is concentrated within a very thin Me-PTC region at the Me-PTC/Au interface, with a field concentration width of d (Fig. 3.7b)

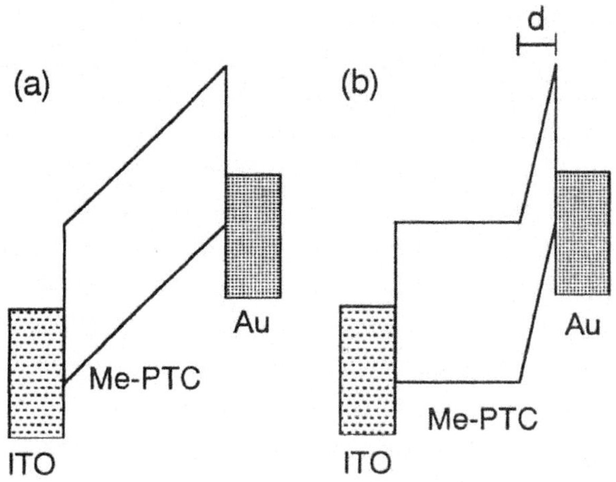

Fig. 3.7 Distribution of electric field application across the Me-PTC film. **a** The applied electric field is assumed to be uniformly distributed throughout the Me-PTC film. **b** The applied electric field is concentrated within a very thin region having a width of d adjacent to the Me-PTC/Au interface

[6, 7]. Then, for various d values, the Φ_B values were estimated using FN analysis. Figure 3.8 shows the relationship between Φ_B and d. It is obvious that as d decreases, i.e., the electric field at the interface becomes more concentrated, Φ_B increases. From this relation, the d value corresponding to a real Φ_B value of 1.50 eV for the Au electrode, which was obtained by photoelectron emission measurements, was estimated as 4.0 nm (Fig. 3.8, curve a).

The same analysis was performed for single-layered cells of Me-PTC films having various thicknesses [6]. Figure 3.9 shows the dependence of d on the thickness of the Me-PTC film. Although the total thickness varied from 200 to 700 nm, d hardly changed. This indicates that a field concentration width of 4 nm represents the condition of the Me-PTC/Au interface during multiplication with considerable accuracy.

Even when Ag electrodes with different barrier heights ($\Phi_B = 0.85$ eV) were used instead of Au ($\Phi_B = 1.50$ eV), the same d value of 4.0 nm was obtained (Fig. 3.8, curve b) [6]. This result further supports the accuracy of the present FN analysis.

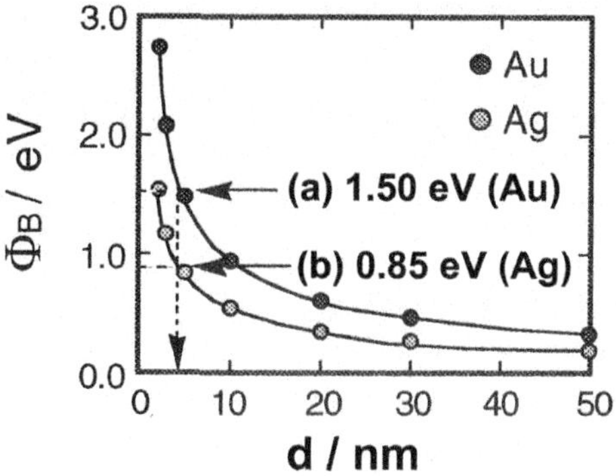

Fig. 3.8 Relationships between the d and Φ_B values estimated from the slopes of the FN plots. The electric field was concentrated around the negatively biased Me-PTC/metal interfaces. Au ($\Phi_B = 1.50$ eV) (**a**) and Ag ($\Phi_B = 0.85$ eV) (**b**) were used as metal electrodes

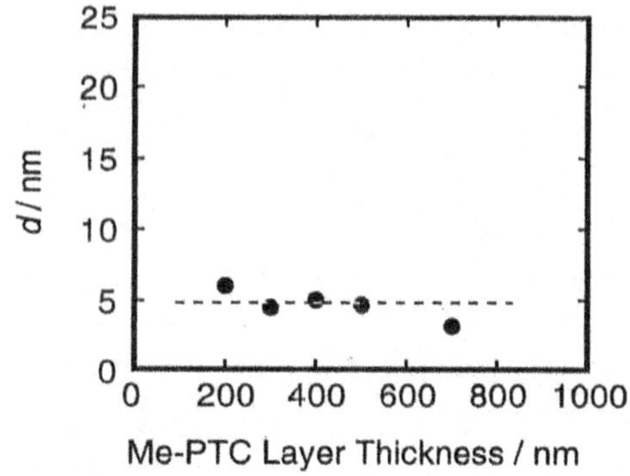

Fig. 3.9 Dependence of d on the thickness of the Me-PTC film

Moreover, this suggests that the electric field concentration width (d) of 4.0 nm is determined by the characteristics of the Me-PTC film rather than those of the metal film.

3.3.3 Estimation of Trap Density

Based on this result, the energy structure of the Me-PTC/Au interface during multiplication is shown in Fig. 3.10a. The organic/metal interface is considered to be subject to violent conditions during multiplication. Assuming that the relative dielectric constant for Me-PTC is 4, an electric field concentration width of 4 nm means that a field strength of 10^7 Vcm^{-1} is built up at the interface. The required number of trapped holes for such an intense field concentration is estimated to be 10^{13} cm^{-2}. This means that 10 holes are trapped in an area of 10 nm × 10 nm (Fig. 3.10b). Considering the molecular size of Me-PTC (approximately 1 nm), this surprisingly high areal density implies a hole trap of some peculiar type [7].

3.4 TSC Measurements

3.4.1 Energetic Depth of Traps

To clarify the nature of these traps, thermally stimulated current (TSC) measurements were performed. First, 600 nm monochromatic light was shone onto the sample at −80 °C to capture and freeze the carriers. Next, the sample was heated at a rate of β°C/min in the dark by applying a carrier collection field of V_c to measure the current due to the release of trapped carriers.

Figure 3.11a shows the TSC curves for a collection field (V_c) of 0.9 V. The Au electrode was negatively biased with respect to the ITO electrode. The TSC peak

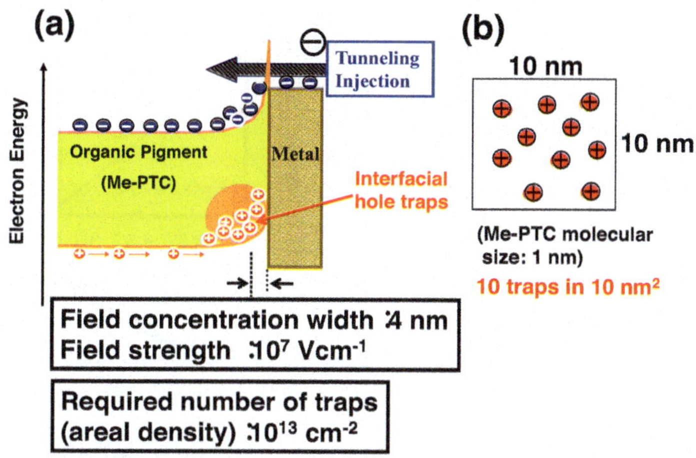

Fig. 3.10 **a** Re-illustrated energy structure of the Me-PTC/Au interface during multiplication. **b** 10 holes were trapped in a 10 nm × 10 nm area. Assuming that the molecular size of Me-PTC is approximately 1 nm, this surprisingly high areal density implies a hole trap of some peculiar type

(T_m) shifted to a higher temperature when the heating rate (β) was increased. If the TSC peak arose merely from the holes released from traps, the energetic depth of the trap (ΔE) could be directly determined from the slope of the $\ln(\beta T_m^{-2})$ versus T_m^{-1} plot, which corresponds to a value of 2ΔEk (k: Boltzmann constant) [13]. However, the very large magnitude of the TSC observed (>10 μAcm^{-2}) indicates that the peak cannot be explained simply by the released holes. Thus, we conclude that the TSC peak observed in the present study is the result of current multiplication due to electron tunneling from the Au electrode caused by hole accumulation at the interfacial traps. This multiplication effect was barely suppressed during the TSC measurements. In this case, the decay of the TSC curves can be attributed to the release of holes from interfacial traps because the reduction in the number of trapped holes results in a smaller number of injected electrons. Moreover, the point of inflection (T_i) of the decay, which coincides with the peak of the differentiated TSC curves, can be regarded as the peak temperature for hole release. Based on the above consideration, $\ln(\beta T_i^{-2})$ was plotted against T_i^{-1} (Fig. 3.11a, inset) and a linear relationship was observed, which was even more linear than that in the case of T_m. Consequently, a ΔE of 0.29 eV was obtained from the slope of the obtained line.

3.4.2 Trap Depth Versus Electric Field

Figure 3.11b shows the dependence of the energetic depth of the trap (ΔE) on the collecting voltage (V_c). Interestingly, ΔE increased with the applied electric field and

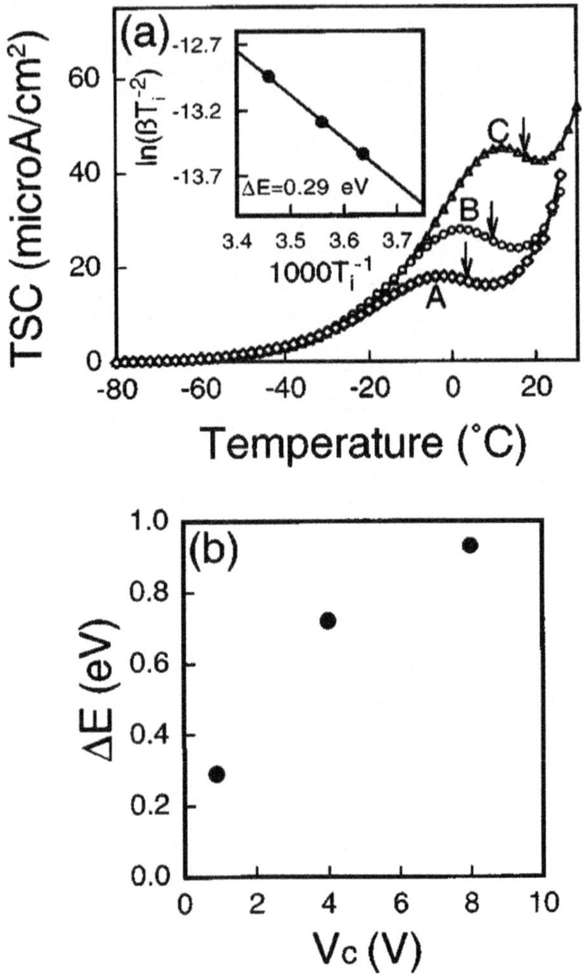

Fig. 3.11 **a** TSC curves of interfacial hole trap observed when applying a V_c of 0.9 V. The Au electrode was biased negatively with respect to the ITO electrode. The heating rates were 6 (curve A), 8 (curve B), and 12 °C/min (curve C). The arrows indicate T_i. Inset is a plot of $\ln(\beta T_i^{-2})$ versus Ti^{-1}. **b** Dependence of the energetic depth of trap (ΔE) on the collecting voltage (V_c). Reproduced with the permission of [7]. © 1998 American Institute of Physics

approached 1 eV at 8.0 V. In the case of an electronic trap-like impurity, ΔE should decrease because of the Poole–Frenkel effect [14–16]. Moreover, it is difficult to imagine hole accumulation in such traps under intense field concentration. Therefore, the author concluded that another type of trap should be considered.

3.5 Structural Trap Model

3.5.1 Experimental Findings

Carrier traps that cause photocurrent multiplication have two distinct characteristics. (1) An extremely high areal density of upto 5 traps per 10 nm^2, which suggests that these traps are of a peculiar type (Fig. 3.10b). (2) An increase in the energetic trap depth with increasing applied electric field (Fig. 3.11b), which is contrary to the behavior of usual electronic traps, i.e., a decrease in trap depth with increasing applied electric field owing to the Poole–Frenkel effect. These anomalous behaviors strongly stimulated the author.

3.5.2 Model Proposal

The author proposes a structural trap model. Figure 3.12 shows a schematic view of the pigment/metal interface. The Me-PTC film has a microscopically rough surface; hence, the metal does not have uniform contact and there are a vast number of sites forming dead ends, which act as hole traps when the field is oriented towards the metal. Assuming a dead end of 1 nm depth under a concentrated electric field of 10^7 Vcm^{-1}, the energy needed to release holes against the field, namely, the energetic depth of the trap (ΔE), reaches 1 eV. Because holes become harder to release, i.e., ΔE increases under a stronger electric field, the TSC result can be reasonably explained by this model.

3.6 Clues for the Origin of Structural Traps

3.6.1 Deposition Rate

At that time, the origin of the structural traps was not known. However, a high areal density reaching 10 traps per 10 nm^2 and dead ends about 1 nm deep implied that the origin was related to the molecular-sized roughness of the surface. This suggests that the multiplication behavior is significantly affected by the organic film crystallinity, which is sensitive to the film preparation conditions. Usually, Me-PTC films (500 nm) are deposited at a rate of 0.2 nms^{-1} (Fig. 3.13, curve A). The author tried to perform high-rate deposition of Me-PTC at upto 2.5 nms^{-1} only for the interfacial region adjacent to the Au electrode in order to reduce the Me-PTC crystallinity at the Me-PTC/Au interface. This led to suppression of the multiplication rate (Fig. 3.13, curve B). This observation suggests a close relationship between molecular stacking and

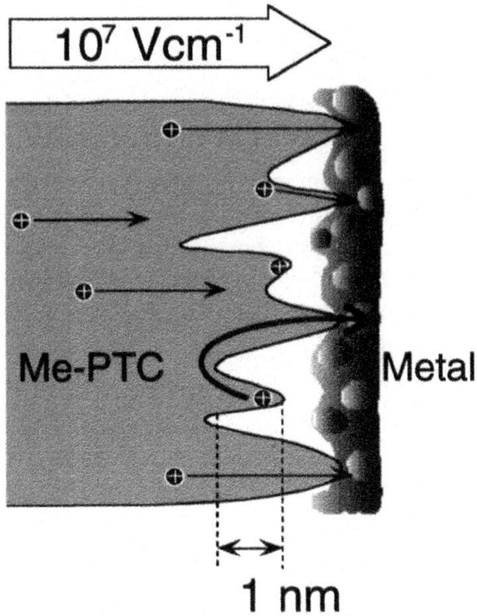

Fig. 3.12 Schematic view of the pigment/metal interface. Holes are trapped by dead ends (structural trap) when the electric field is oriented towards the metal. Reproduced with the permission of [7]. © 1998 American Institute of Physics

the formation of structural traps, and strongly supports the existence of structural traps.

3.6.2 Material Dependence

FN analysis was performed for various organic semiconductors other than Me-PTC (Figs. 3.6 and 3.8), such as NTCDA (Fig. 3.14a) and t-BuPh-PTC (Fig. 3.15a). The thicknesses of all organic semiconductor films were the same (500 nm).

NTCDA obeyed the FN relationship (Fig. 3.14b) at the high electric field strength at which large multiplication occurred regardless of whether Au or Ag electrodes were used. The same field concentration width (d) of 2 nm was obtained for both Au and Ag electrodes (Fig. 3.14c) [17]. t-Bu-PTC also obeyed the FN relationship under multiplication in a high electric field (Fig. 3.15b), and a field concentration width (d) of 10 nm was obtained (Fig. 3.15c) [18].

As shown in Table 3.1, as semiconductors with higher multiplied photocurrent densities were used, in the order of t-BuPh-PTC, Me-PTC, and NTCDA, the electric field concentration width (d) decreased in the order of 10, 4, and 2 nm. This clearly indicates that a higher multiplication rate requires a stronger field concentration at the pigment/Au interface, which is formed by a larger areal density of trapped holes.

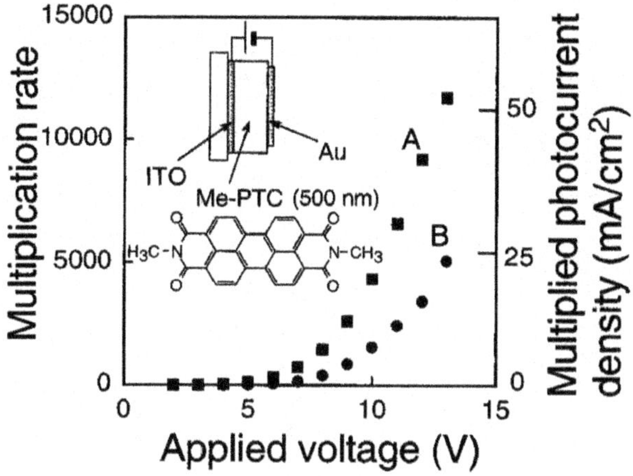

Fig. 3.13 Dependence of the multiplication rate and multiplied photocurrent density on the applied voltage for the cells when the entire Me-PTC film was deposited at a rate of 0.2 nms^{-1} (curve A) and when only the interfacial region of the pigment (50 nm) in the vicinity of the Au electrode was deposited at a high rate of 2.5 nms^{-1} (curve B). The Au electrode was irradiated with monochromatic light (600 nm, 0.04 $mWcm^{-2}$). Reproduced with the permission of [7]. © 1998 American Institute of Physics

The increase in the areal density of traps is expected to be caused by an increase in the molecular roughness in the order of t-BuPh-PTC, Me-PTC, and NTCDA. Because t-BuPh-PTC has bulky side chains, the crystallinity of the deposited film was low. In contrast, because NTCDA has no side chains, the crystallinity of the deposited film was high. This observation also suggests that the structural trap is related to the crystallinity of the organic semiconductor films. Detailed research on clarifying the origin of the structural trap is described in Chaps. 4 and 5.

3.6.3 Relationship Between Trap Density and Onset Charge

In Sect. 3.2.5, based on the transient response measurements, the minimum hole charge required to cause the tunneling injection of electrons (onset charge) was determined to be 30 $nCcm^{-2}$, which corresponds to an areal density of 2×10^{11} cm^{-2} [4]. The onset time and time when sufficient photocurrent multiplication occurs are approximately several tens of milliseconds and several seconds, respectively, i.e., the photocurrent multiplication time is approximately 100 times longer than the onset time. Thus, the areal density of trapped holes that can cause sufficient tunneling injection of electrons can be estimated to be approximately 2×10^{13} cm^{-2} by assuming that the capture probability of photogenerated holes by traps at the Me-PTC/Au interface is close to unity. The density of trapped holes, i.e., the number of holes passing through the Me-PTC/Au interface, should have a close relationship with

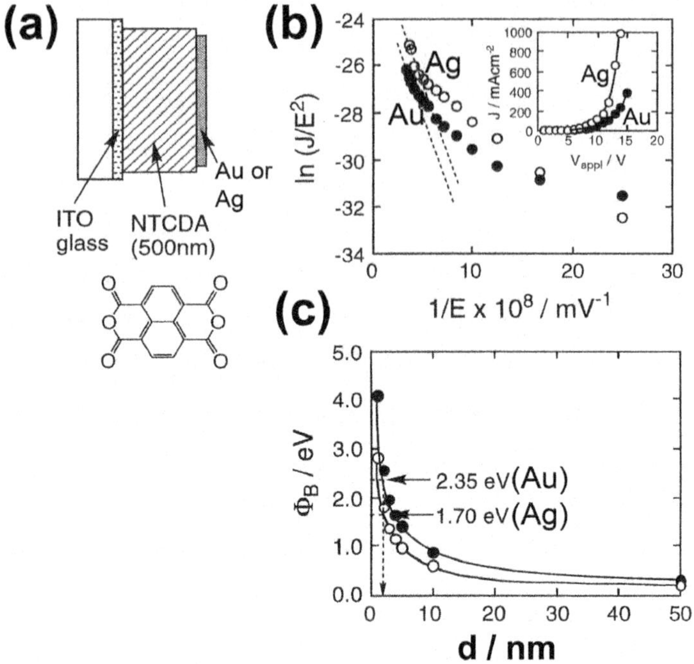

Fig. 3.14 **a** Cell structure using the NTCDA film. **b** FN plots for the NTCDA/Ag and NTCDA/Au interfaces. Inset: Multiplied photocurrent density (J) versus applied voltage (V_{appl}) plot. Au or Ag electrodes were biased negatively with respect to ITO electrode. **c** Relationship between the barrier height for electron injection (Φ_B) and field concentration width (d). $\Phi_B = 2.35$ eV for Au and $\Phi_B = 1.70$ eV for Ag obtained by atmospheric photoelectron spectroscopy were used to estimate d

the value estimated by the FN plots, i.e., the number of holes captured by the structural trap (Sect. 3.3.3) [7]. Detailed investigations combining the FN analyses [6, 7] and transient response measurements [4, 5] may lead to a more precise understanding of the photocurrent multiplication process, especially for aspects of the hole capturing process of structural traps, such as the capture probability.

3.7 Conclusions

Carrier traps at the perylene pigment/metal interface, which give rise to a large photocurrent multiplication phenomenon with more than 10^4-fold multiplication, were investigated. The areal density of the traps was estimated to be 10 traps per 10 nm^2 by FN analysis. The energetic depth of the traps increased with the applied electric field, contrary to the behavior of usual electronic traps. To explain these anomalous behaviors, a model of molecular-sized structural traps was proposed. The model posits that dead ends at imperfect organic/metal contacts capture holes more

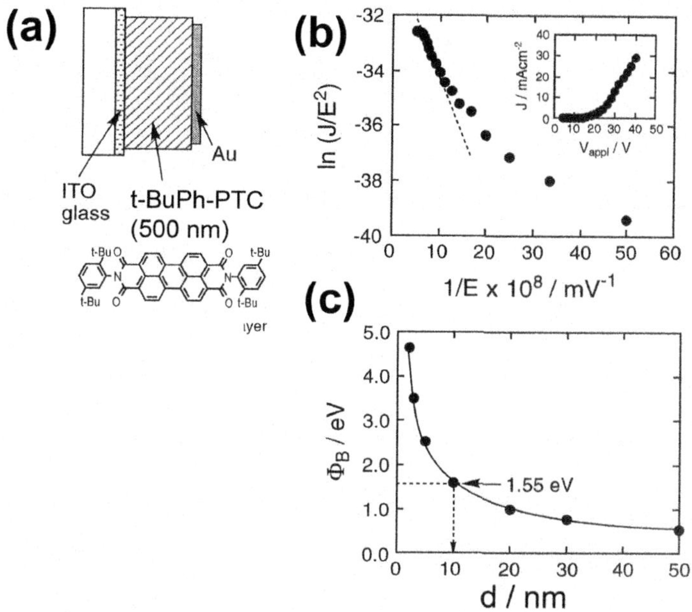

Fig. 3.15 **a** Cell structure using the t-BuPh-PTC film. **b** FN plots for the t-BuPh-PTC/Ag and t-BuPh-PTC/Au interfaces. Inset: Multiplied photocurrent density (J) versus applied voltage (V_{appl}) plot. The Au electrode was biased negatively with respect to ITO electrode. **c** Relationship between Φ_B and d. $\Phi_B = 1.55$ eV obtained by atmospheric photoelectron spectroscopy was used to estimate d

Table 3.1 Electric field concentration width (d) obtained by FN plots and the maximum multiplied photocurrent density observed for the NTCDA/Au, Me-PTC/Au, and t-BuPh-PTC/Au interfaces. Voltages corresponding to the maximum multiplied photocurrent density are indicated in the brackets

Organic semiconductor film	Field concentration width (d) (nm)	Maximum multiplied photocurrent density
NTCDA	2	400 $mAcm^{-2}$ (15 V)
Me-PTC	4	125 $mAcm^{-2}$ (22 V)
t-BuPh-PTC	10	30 $mAcm^{-2}$ (40 V)

tightly under a stronger electric field. The estimated value of 10 trapped holes per 10 nm^2 and dead ends of the order of 1 nm suggest that the proposed structural traps are closely related to the molecular-sized roughness of the pigment surface.

References

1. Hiramoto, M., Imahigashi, T., Yokoyama, M.: Photocurrent multiplication in organic pigment films. Appl. Phys. Lett. **64**, 187–189 (1994)
2. Hiramoto, M., Kawase, S., Yokoyama, M.: Photoinduced hole injection multiplication in p-type quinacridone pigment films. Jpn. J. Appl. Phys. **35**, L349–L351 (1996)
3. Katsume, T., Hiramoto, M., Yokoyama, M.: Photocurrent multiplication in naphthalene tetracarboxylic anhydride film at room temperature. Appl. Phys. Lett. **69**, 3722–3724 (1996)
4. Nakayama, K., Hiramoto, M., Yokoyama, M.: Direct tracing of the photocurrent multiplication process in an organic pigment film. J. Appl. Phys. **84**, 6154–6156 (1998)
5. Nakayama, K., Hiramoto, M., Yokoyama, M.: Transient response of multiplied photocurrent observed in metal/organic pigment film interface. In: Proceedings of IS&T's NIP14: International Conference on Digital Printing Technologies, Toronto, Ontario, Canada, pp. 490–493, 18–23 Oct (1998)
6. Katsume, T., Hiramoto, M., Yokoyama, M.: Detailed analysis of photocurrent multiplication phenomenon in organic pigment films. In: Proceedings of IS&T's NIP12: International Conference of Digital Printing Technologies, San Antonio, Texas, pp. 248–251, Oct 27-Nov. 1 (1996)
7. Hiramoto, M., Nakayama, K., Katsume, T., Yokoyama, M.: Field-activated structural traps at organic pigment/metal interfaces causing photocurrent multiplication phenomena. Appl. Phys. Lett. **73**, 2627–2629 (1998)
8. Fowler, R.H., Nordheim, L.: Electron emission in intense electric fields. Proc. R. Soc. London Ser. A **119**, 173–181 (1928)
9. Simmons, J.O.: Generalized formula for the electric tunnel effect between similar electrodes separated by a thin insulating film. Appl. Phys. Lett. **34**, 1793–1803 (1963)
10. Parker, I.D.: Carrier tunneling and device characteristics in polymer light-emitting diodes. J. Appl. Phys. **75**, 1656–1666 (1994)
11. Sze, S.M.: Physics of Semiconductor Devices, Wiley, New York, p. 552 (1981)
12. Kirihata, H., Uda, M.: Externally quenched air counter for lowenergy electron emission measurements. Rev. Sci. Instrum. **52**, 68–70 (1981)
13. Booth, A.H.: Calculation of electron trap depths from thermoluminescence maxima. Can. J. Chem. **32**, 214–215 (1954)
14. Frenkel, J.: On pre-breakdown phenomena in insulators and electronic semiconductors. Phys. Rev. **54**, 647–648 (1938)
15. Simmons, J.G.: Poole-Frenkel effect and Schottky effect in metal-insulator-metal systems. Phys. Rev. **155**, 657–660 (1967)
16. Harrell, W.R., Frey, J.: Observation of Poole-Frenkel effect saturation in SiO_2 and other insulating films. Thin Solid Films **352**, 195–204 (1999)
17. Unpublished results
18. Unpublished results

Chapter 4
Morphology of Organic/Metal Interface and Photocurrent Multiplication Behaviors

4.1 Photocurrent Multiplication and Structural Trap Model

We observed a large 10^4-fold photocurrent multiplication in various organic semiconductor films [1–4] (Chap. 2). A mechanism based on the tunneling injection of electrons from the metal electrode to the pigment layer under a high electric field built up by the trapped holes near the organic/metal interface was proposed (Fig. 4.1a) [1–4]. The carrier trap at the organic/metal interface that causes multiplication was revealed to exhibit abnormal characteristics (Chap. 3), including (i) an extremely high areal density reaching 10 traps per 10 nm^2 and (ii) an increase in the energetic trap depth with an increase in the applied electric field [5–8]. A structural trap model was proposed (Fig. 4.1b) [8] in which the blind alleys at an imperfect organic/metal contact capture held tightly under a strong electric field (Chap. 3). This model strongly suggested a close relationship between the morphology of the organic/metal interface and the multiplication behaviors.

In this chapter, the influence of the morphologies of organic films and multiplication behaviors and those of metal films and multiplication behaviors are described in detail. To approach the true nature of the structural trap formed at the organic/metal interface, a vast number of experiments, including trial and error, are required. Finally, a revised structural trap model was proposed.

M. Hiramoto, *Organic Photocurrent Multiplication*, Electronic Materials: Science & Technology, https://doi.org/10.1007/978-981-99-1237-7_4

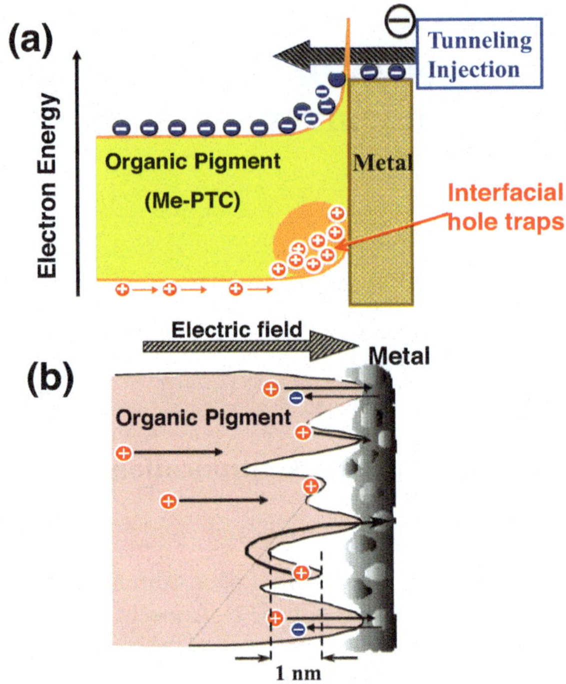

Fig. 4.1 **a** Energy structure of Me-PTC/Au interface during multiplication. **b** Schematic illustration of structural traps formed at the organic/metal interface. Holes were trapped by blind alleys when the electric field was toward the metal. Reproduced with the permission of [8] and [9]. © 1998 American Institute of Physics

4.2 Morphology of Organic Films and Multiplication Behaviors

4.2.1 Motivation

The structural trap model [8–11] implied that the multiplication behaviors were extremely sensitive to the surface morphology of the organic films. Thus, the influence of the surface morphology of organic films on the multiplication behaviors was investigated to confirm the existence of a structural trap (Fig. 4.1b).

4.2.2 Three Perylene Pigments with Different Side Chains

Three derivatives of perylene pigments (PTC) with different side chains, i.e., Me-PTC, n-Bu-PTC, and PhEt-PTC (Fig. 4.2), were used. The cells using these pigments were fabricated under the same fabrication conditions. The PTCs were purified twice by train sublimation [12]. Sandwich-type cells (Fig. 4.2) were fabricated by vacuum evaporation. The PTCs and Au were deposited on an indium tin oxide (ITO) glass

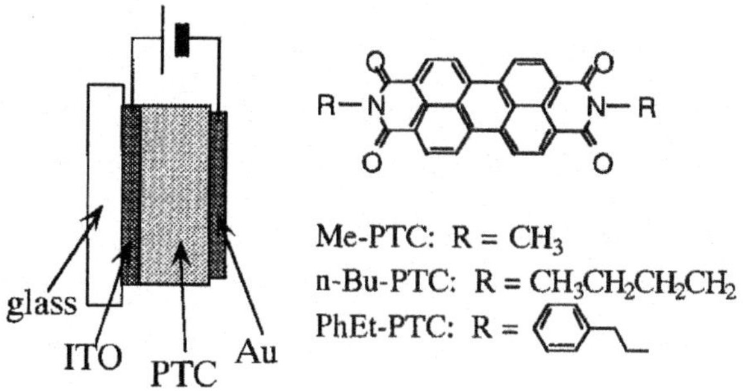

Fig. 4.2 Cell structure and chemical formulas of the perylene pigments used. Reproduced with the permission of [9]. © 1000 American Institute of Physics

substrate under 1×10^{-3} Pa at room temperature. The deposition rate for the PTCs was adjusted to 0.2 nms^{-1}. The deposition rate of Au was maintained at 0.04 nms^{-1}.

4.2.3 *Multiplication Rates*

Figure 4.3 shows the dependence of the multiplication rate on the applied voltage for the sandwich cells (Fig. 4.2) incorporating Me-PTC (curve a), Bu-PTC (curve b), and PhEt-PTC (curve c) under the same measuring conditions. The multiplication behavior was strongly dependent on the type of the PTC pigments. Me-PTC and n-Bu-PTC exhibited large multiplication rates exceeding 2×10^4 and 5×10^3 at 20 V, respectively. Interestingly, PhEt-PTC hardly showed multiplication, i.e., its multiplication rate was only 1.9 even at 20 V.

4.2.4 *Morphology of Organic Films*

Figure 4.4 shows the scanning electron microscopy (SEM) images of the as-deposited films of Me-PTC (a), n-Bu-PTC (b), and PhEt-PTC (c). For Me-PTC and n-Bu-PTC with relatively small substituents, the deposited films were observed to contain a gathering of microcrystals and to possess a rough surface, since the molecules tended to form aggregates easily, owing to strong intermolecular interactions. Contrarily, for PhEt-PTC with a bulky substituent of phenylethyl, the deposited film had a considerably flat and smooth surface.

4.2.5 *Crystallinity of Organic Films*

Figure 4.5 shows the X-ray diffraction (XRD) patterns of these organic films. The Me-PTC film (a) exhibited intense diffraction peaks. However, the PhEt-PTC film

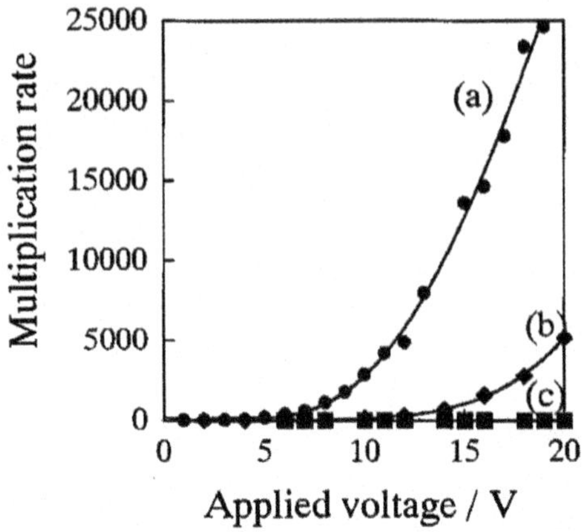

Fig. 4.3 Dependence of multiplication rate on the applied voltage for the ITO/PTC(500 nm)/Au(20 nm) cells. Me-PTC (curve a), n-Bu-PTC (curve b), and PhEt-PTC (curve c) were used as organic films. The Au electrode was negatively biased with respect to the ITO electrode. Monochromatic light of 600 nm (0.04 mW cm^{-2}) was irradiated on the Au electrode. Measurements were performed at room temperature. Reproduced with the permission of [9]. © 1000 American Institute of Physics

(b) did not show diffraction peaks. Thus, it was concluded that the Me-PTC and n-Bu-PTC films with large multiplication were polycrystalline, and the PhEt-PTC film without multiplication was amorphous.

4.2.6 Multiplication Induction by Crystallization

The as-deposited PhEt-PTC film is known to easily undergo a phase change from amorphousness to crystallinity upon exposure to tetrahydrofuran (THF) vapor [13, 14]. The SEM images of the as-deposited and THF-exposed PhEt-PTC films are shown in Fig. 4.4c, d, respectively. The smooth and flat surface of the as-deposited film became rough and resembled a gathering of bamboo leaf-like crystals several micrometers long after the THF exposure. The XRD patterns indicated that the as-deposited and THF-exposed films showed no peak and a sharp diffraction peak at 6° [14], indicating that amorphousness and crystallinity, respectively (Fig. 4.5b, c).

Figure 4.6 shows the dependence of the multiplication rate on the applied voltage for the ITO/PhEt-PTC/Au cells. When the as-deposited film was used (curve a), the multiplication rate was 1.2 under a bias of 16 V. Interestingly, when the THF-exposed film was used (curve b), the multiplication rate was 360-fold at 18 V. The photocurrent multiplication was induced by changing the crystallinity of the organic film from amorphousness to polycrystallinity.

This result indicated that the organic film should be crystallized for photocurrent multiplication.

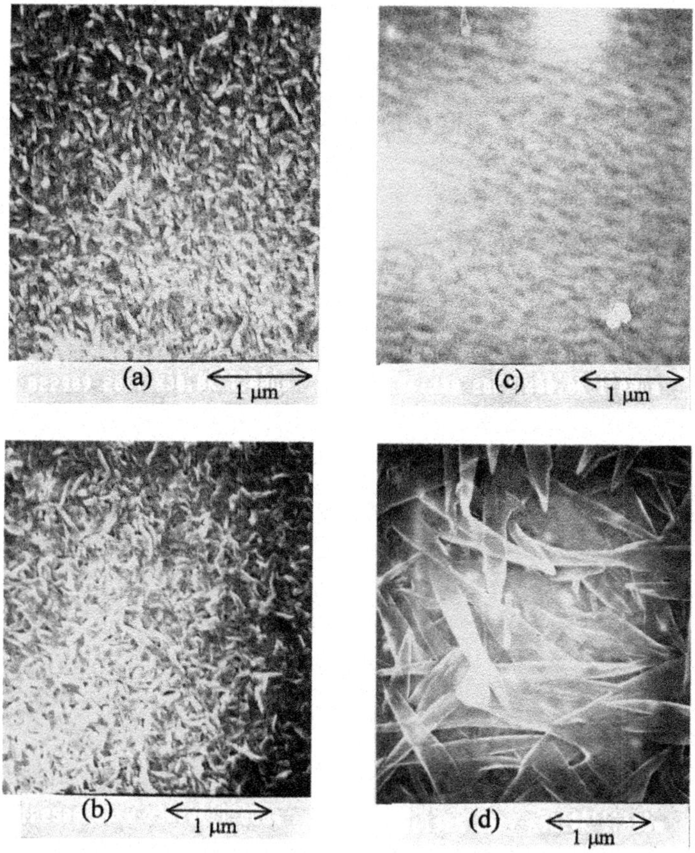

Fig. 4.4 Scanning electron microscopy (SEM) images for the as-deposited films of Me-PTC (**a**), n-Bu-PTC (**b**), PhEt-PTC (**c**), and THF-exposed film of PhEt-PTC (**d**). The thicknesses of the films were 500 nm. Reproduced with the permission of [9]. © 1000 American Institute of Physics

4.2.7 Energetic Barrier Height at Organic/Metal Junctions

The multiplication mechanism shown in Fig. 4.1a implied that the multiplication behavior depended on the barrier height for electron injection from the Au to the conduction state of the PTC films. Figure 4.7 shows the energy diagrams of Me-PTC, n-Bu-PTC, and PhEt-PTC, with the work function of the Au film. The barrier heights did not range in the order of the multiplication rate, and the Me-PTC film with the largest multiplication rate had the highest barrier. This result indicated that the energetic difference of the perylene pigments cannot explain the large difference in the multiplication rate from unity to 2×10^4-fold.

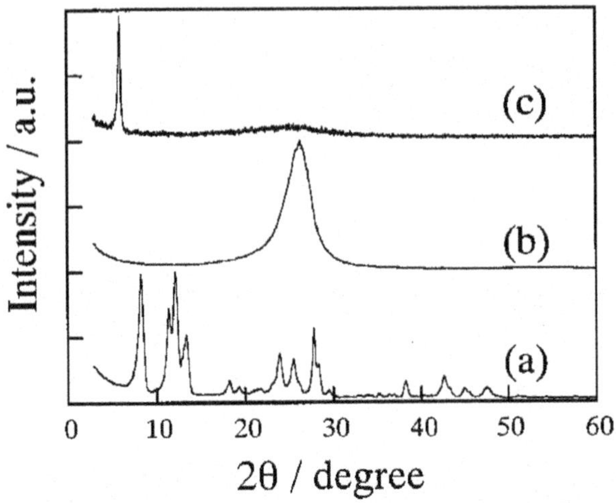

Fig. 4.5 X-ray diffraction (XRD) patterns for the PTC films (500 nm). **a** Me-PTC (as-deposited), **b** PhEt-PTC (as-deposited), and **c** PhEt-PTC (after THF vapor treatment). Reproduced with the permission of [9]. © 1000 American Institute of Physics

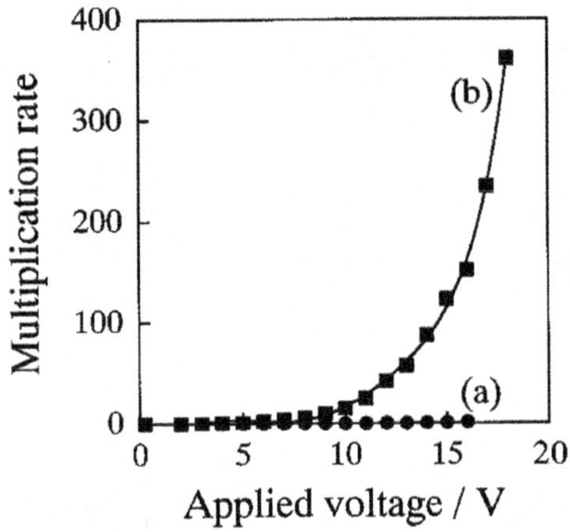

Fig. 4.6 Dependence of multiplication rate on the applied voltage for the ITO/PhEt-PTC(500 nm)/Au(20 nm) cells. The as-deposited (curve a) and THF-exposed (curve b) PhEt-PTC films were used. The Au electrode was negatively biased with respect to the ITO electrode. Measurements were performed at 233 K. Monochromatic light of 600 nm (0.04 mW cm^{-2}) was irradiated on the Au electrode. Reproduced with the permission of [9]. © 1000 American Institute of Physics

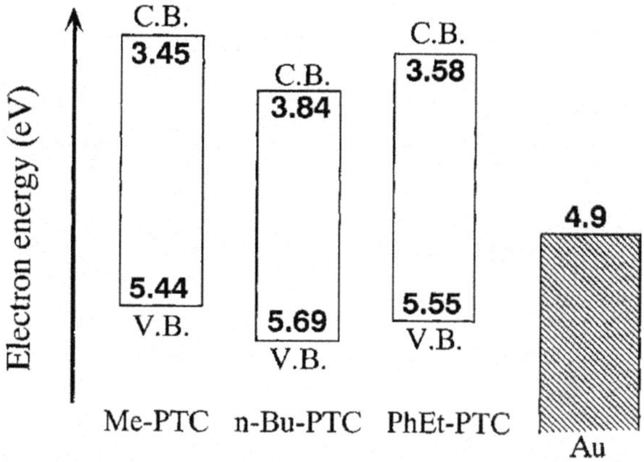

Fig. 4.7 Energy diagrams of the Me-PTC, n-Bu-PTC, and PhEt-PTC, and Au films. VB and CB denote the valence band and conduction band, respectively. The work function of the Au electrode and the energy value of the upper edges of the VB of the PTC films from the vacuum level (Me-PTC, 5.44 eV; n-Bu-PTC, 5.69 eV; and PhEt-PTC, 5.55 eV) was measured by atmospheric photoelectron emission analysis [12]. The optical bandgaps of Me-PTC (1.99 eV), n-Bu-PTC (1.85 eV), and PhEt-PTC (1.97 eV) were estimated from the edge of the absorption spectra. From these values, the lower edges of the CB and the barrier heights of Me-PTC (1.45 eV), n-Bu-PTC (1.06 eV), and PhEt-PTC (1.32 eV) were obtained. Reproduced with the permission of [9]. © 1000 American Institute of Physics

4.2.8 Explanation Based on Structural Trap Model

The structural trap model can reasonably explain the observed results. The surface of polycrystalline PTC films such as Me-PTC, n-Bu-PTC, and THF-exposed PhEt-PTC can be assumed to be microscopically rough, whereas that in the amorphous state, such as the as-deposited PhEt-PTC film, was assumed to be microscopically smooth. Therefore, for the crystalline films, a large number of blind alleys at which holes could not pass through the organic/metal interface may have existed, owing to uneven contact. The trapped carriers captured at the organic/metal interface caused an effective field concentration, leading to a large multiplication rate (Fig. 4.8a). However, for the amorphous films, there were a few blind alleys. Moreover, even if there were spatial gaps between the organic and metal, the hole easily moved into the metal through a smooth organic surface. As a result, multiplication hardly occurred (Fig. 4.8b). Considering the strong effect of crystallization on the multiplication behavior, there should be a close relationship between molecular stacking and the formation of blind alleys.

4.2.9 Lateral Spatial Separation of Trap Sites and Injection Sites

The spatial structure of the organic/metal interface with a structural trap (Fig. 4.8) was characterized by two types of energetic structures. The first was an organic/metal

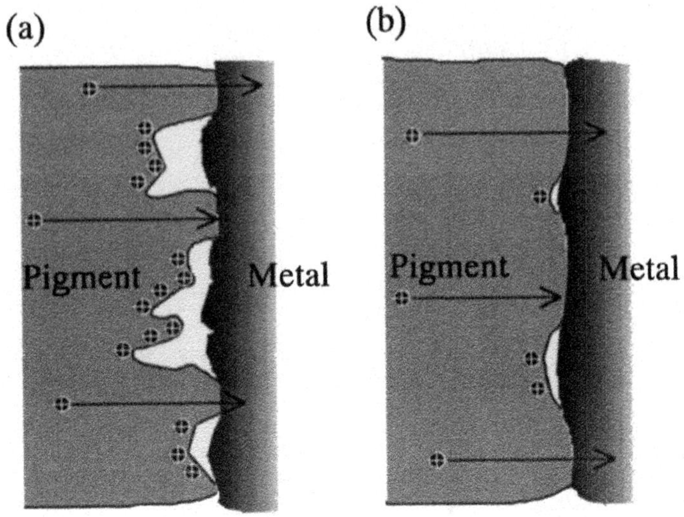

Fig. 4.8 Structural trap model. The schematic view of the organic/metal interface is illustrated. **a** Crystalline PTC films with a rough surface. Holes were trapped by blind alleys at the pigment/metal interface when the electric field was toward the metal. **b** Amorphous PTC films with a smooth surface. Holes easily passed through the pigment/metal interface. Reproduced with the permission of [9]. © 1000 American Institute of Physics

junction at the interface, where the organic and metal films directly contacted each other. The second was an organic/vacuum/metal junction, where there were spatial gaps. For the latter, a vacuum level was inserted in the organic/metal interface. Based on this spatially intermingled energetic model, the electron injection from the metal electrode only occurred at the contacted part. Figure 4.1a shows the energy diagram of only the contacted part. At the noncontacted parts, the barrier height to vacuum level was extremely high, and the barrier width was too wide to prevent tunneling; electrons could not be injected, and trapped holes could not escape. Considering that both parts were spatially separated, the trapped holes can prevent recombination with the injected electrons and form a high electric field for electron injection.

4.2.10 Identification of Interfacial Crystallization Effect

The entire PhEt-PTC film, including the bulk, was crystallized via THF treatment. Similarly, for the as-deposited films of Me-PTC, n-Bu-PTC, and PhEt-PTC, the morphology and crystallinity of the films, including the surface and bulk, differed depending on the substituent type. However, based on the structural trap model, only the molecular stacking near the organic/metal interface should affect the multiplication behavior. To confirm this, we attempted to disturb the molecular stacking of only the neighboring parts of the organic/metal interface by high-rate evaporation.

Figure 4.9 shows the dependence of the multiplication rate on the applied voltage for the ITO/n-Bu-PTC(500 nm)/Au cells. The cell structures are shown in the figure. When the organic film was deposited at a slow rate of 0.2 nms^{-1}, large multiplication was observed (curve a). When the entire organic film was deposited at an overly high rate of 3.0 nms^{-1}, the multiplication was effectively suppressed because the high deposition rate inhibited crystal formation (curve b). Interestingly, when only an extremely thin organic layer (thickness, 50 nm) neighboring the Au electrode was evaporated at an exceptionally high deposition rate of 3.0 nms^{-1}, a suppressed multiplication rate similar to that shown in curve b was observed (curve c). Fundamentally, the same result was obtained for the Me-PTC film (Chap. 3) [8]. These results showed that multiplication was suppressed only by disturbing the molecular stacking near the organic/metal interface, as expected from the structural trap model.

A close relationship between the multiplication behaviors and organic morphology, precisely between the multiplication behaviors and organic crystallinity, was observed. This result strongly supported the existence of a structural trap.

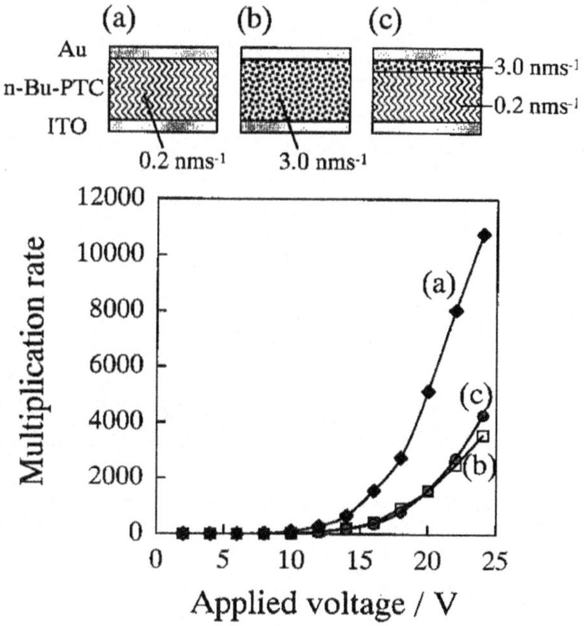

Fig. 4.9 Dependence of multiplication rate on the applied voltage for the ITO/n-Bu-PTC(500 nm)/Au cells. **a** n-Bu-PTC was deposited at a rate of 0.2 nms^{-1}. **b** n-Bu-PTC was deposited at a rate of 3.0 nms^{-1}. **c** The 450 nm-thick n-Bu-PTC film was deposited at a rate of 0.2 nms^{-1}, and the residual 50-nm-thick n-Bu-PTC film was deposited at a rate of 3.0 nms^{-1}. The Au electrode was negatively biased with respect to the ITO electrode. Reproduced with the permission of [9]. © 1000 American Institute of Physics

4.3 Metal Morphology Observed by SEM and Multiplication Behaviors

4.3.1 Motivation

The structural trap model (Fig. 4.1b) [8–11] implied that the morphology of the metal films influenced the photocurrent multiplication behaviors. Thus, the morphology of the metal films deposited on organic films under various preparation conditions was observed, and the relationship between the metal morphology and photocurrent multiplication behaviors was clarified.

4.3.2 Effects of Au Deposition Conditions on Perylene Pigment Film [15]

Sandwich-type cells were fabricated using the Me-PTC film (Fig. 4.10). The deposition conditions for the 20 nm-thick Au film were varied. The deposition conditions of the Me-PTC films were maintained, i.e., the deposition rate was constant at 0.2 nms^{-1} under 1×10^{-3} Pa. The depositions were performed at room temperature for all substrates.

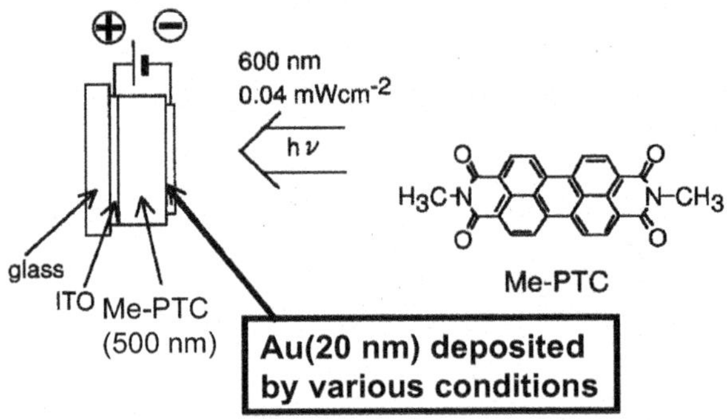

Fig. 4.10 Cell structure and chemical formula of Me-PTC. The deposition conditions for the Au electrode on the Me-PTC film were varied. The deposition conditions of the Me-PTC films were the same. Reproduced with the permission of [14]. © 1998 The Physical Society of Japan and The Japan Society of Applied Physics

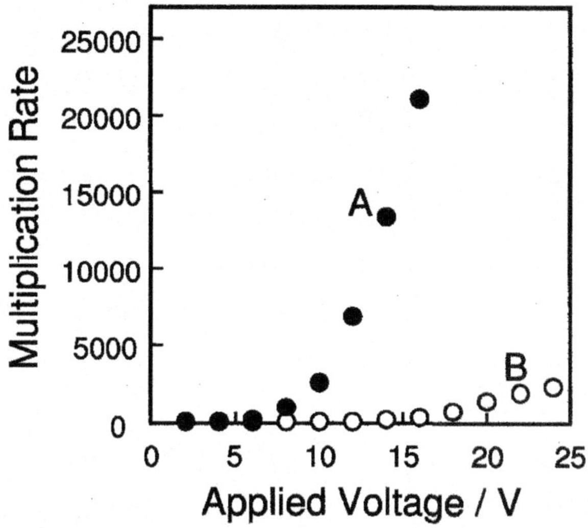

Fig. 4.11 Dependence of multiplication rate on the applied voltage observed for the ITO/Me-PTC/Au cells deposited using two techniques: **a** Au film (20 nm) deposited by vacuum evaporation. **b** Au film (2 nm) deposited by ion sputtering, and the residual Au (18 nm) deposited by vacuum evaporation. Measurements were performed at −50 °C. Reproduced with the permission of [14]. © 1998 The Physical Society of Japan and The Japan Society of Applied Physics

4.3.3 Resistivity-Heated Deposition Versus Sputtering Deposition

Figure 4.11 shows the dependence of the multiplication rate on the applied voltage observed for the ITO/Me-PTC/Au cells prepared using different Au deposition techniques. When the Au film was deposited by vacuum evaporation from a resistivity-heated boat, the multiplication rate exceeded 2×10^4-fold at 16 V (curve A). Contrarily, when a 2 nm-thick Au film was deposited by low-power ion sputtering, and the residual 18-nm-thick Au was deposited by vacuum evaporation, the multiplication rate was drastically suppressed, and a small multiplication rate of 2×10^3-fold was observed at a high voltage of 24 V (curve B) [15].

4.3.4 Deposition Rate

Figure 4.12 shows the dependence of the multiplication rate on the Au deposition rate during vacuum evaporation. The multiplication rate increased 30-fold from 6×10^2 to 1.8×10^4-fold by decreasing the deposition rate from 0.7 to 0.008 nms^{-1}.

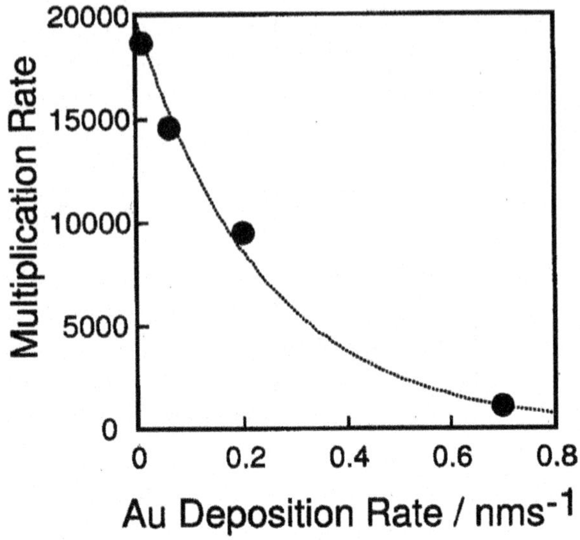

Fig. 4.12 Multiplication rate dependence on the Au deposition rate during vacuum evaporation. Multiplication rates were compared with the values at an applied voltage of 17 V. Reproduced with the permission of [14]. © 1998 The Physical Society of Japan and The Japan Society of Applied Physics

Figures 4.11 and 4.12 indicate that the multiplication behaviors were quite sensitive to the preparation conditions of the Au films [15].

4.3.5 *Morphology of Au Films Observed by SEM*

Figure 4.13 shows the SEM images of the Me-PTC films covered with a 2-nm-thick Au film prepared using different deposition methods [15]. When Au was deposited by low-power ion sputtering, a unique rough surface comprising a cluster of 100-nm nanoscale crystals was observed (Fig. 4.13b). The SEM image of the Me-PTC film without Au practically exhibited the same morphology. Thus, we concluded that the Au film deposited by sputtering fairly adhered uniformly to the microscopically rough surface of the Me-PTC film.

Oppositely, when Au was deposited by vacuum evaporation at an extremely low rate of 0.008 nms^{-1}, only the flat morphology of the deposited Au film appeared to cover the entire polycrystalline surface structure of the Me-PTC film (Fig. 4.13a). Therefore, it was concluded that the Au film with a virtually flat morphology hid the rough Me-PTC surface. For vacuum evaporation at a high rate of 0.7 nms^{-1}; however, a morphology similar to that shown in Fig. 4.13b was observed, i.e., Au did not form aggregations and uniformly adhered to Me-PTC.

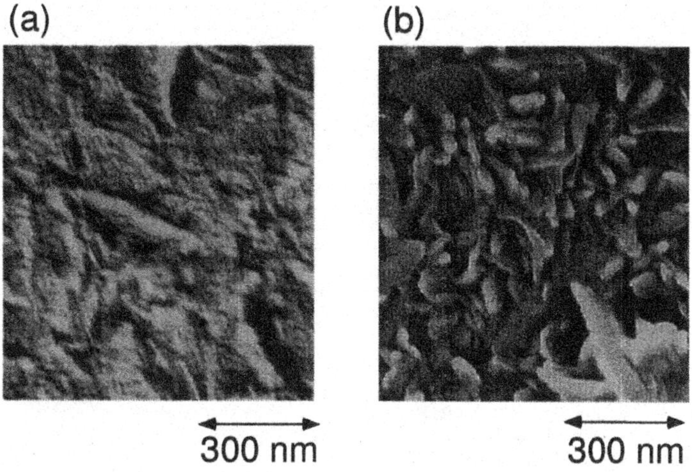

Fig. 4.13 Scanning electron microscopy (SEM) images of the Me-PTC films covered with 2 nm-thick Au films deposited by a low-rate vacuum evaporation at 0.008 nms^{-1} (**a**) and by low-power ion sputtering (**b**). Reproduced with the permission of [14]. © 1998 The Physical Society of Japan and The Japan Society of Applied Physics

4.3.6 Explanation by Structural Trap Model

Summarily, the Me-PTC/Au interface formed by the Au film that flatly aggregated itself exhibited a large photocurrent multiplication, and that formed using the Au film that uniformly adhered to the Me-PTC surface exhibited a small photocurrent multiplication. These two types of organic/metal interfaces are schematically illustrated in Fig. 4.14.

When the deposited Au film flatly aggregated (Fig. 4.14a), it did not uniformly cover the microscopically rough surface of the Me-PTC film and was partially separated from the Me-PTC surface. In this type of organic/metal interface, numerous blind alleys existed, acting as the sites preventing the holes from passing through to the metal electrode. This accumulation of photogenerated holes resulted in a large multiplication.

Contrarily, when the deposited Au adhered to the Me-PTC surface (Fig. 4.13b), the Au film uniformly covered the entire rough surface of the Me-PTC film. In this type of organic/metal interface, there were few blind alleys, holes easily passed through the interface to the Au electrode, and multiplication hardly occurred.

The relationship between the Au morphologies and multiplication behavior was reasonably explained based on the structural trap model.

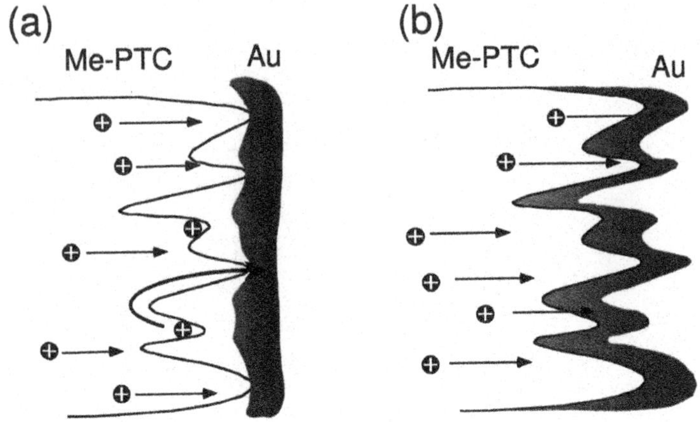

Fig. 4.14 Schematic illustration of the organic/metal interfaces. **a** Au film flatly aggregated itself and was partially separated from the Me-PTC surface. The holes were trapped by blind alleys when there was an electric field toward the metal surface. **b** The Au film uniformly adhered to the Me-PTC surface and covered the entire rough surface of the Me-PTC film. Reproduced with the permission of [14]. © 1998 The Physical Society of Japan and The Japan Society of Applied Physics

4.4 Metal Morphology Observed by AFM and Multiplication Behaviors

4.4.1 Motivation

The relationship between the Au morphology observed by SEM and multiplication behaviors was reasonably explained based on the structural trap model [15]. However, it was strongly considered that a high spatial resolution was required to expose the molecular-level nanostructure of the metal film. Thus, the metal structure was observed by atomic force microscopy (AFM).

4.4.2 Morphology of Au Film on 1, 4, 5, 8-Naphthalenetetracarboxylic Dianhydride (NTCDA) [16]

As an organic/metal interface, we selected NTCDA and Au (Fig. 4.15a). Sandwich-type cells of ITO/NTCDA (400 nm)/Au exhibited photocurrent multiplication exceeding 10^5-fold at 13 V (Fig. 4.15b, curve A) when the Au film was deposited by resistivity-heated vacuum evaporation. The multiplication mechanism at the NTCDA/Au interface was the same as that at the Me-PTC/Au interface (Fig. 4.1a) [1, 3].

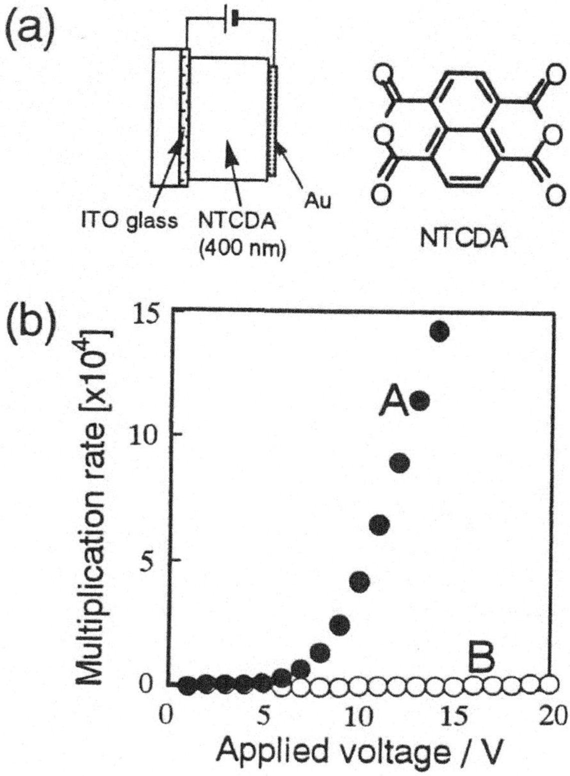

Fig. 4.15 **a** Cell structure and chemical formula of NTCDA. NTCDA (400 nm) was deposited on an ITO glass substrate under 1×10^{-3} Pa. **b** Dependence of multiplication rate on the applied voltage. Au was deposited under 1×10^{-3} Pa from the resistivity-heated tungsten boat (curve A) or the EB evaporation source (curve B). The Au electrode was negatively biased with respect to the ITO electrode. Measurements were performed at room temperature. Reproduced with the permission of [15]. © 2002 IEICE

Figure 4.16 shows the AFM image of the NTCDA film without Au. The NTCDA film (400 nm) exhibited a gathering of relatively large crystals approximately 100 nm long. An intense peak in the XRD pattern (11.68°) indicated that the NTCDA film was polycrystalline. Figure 4.17 shows the AFM images of NTCDA with Au. When Au was deposited on the NTCDA film from a resistively heated tungsten boat, the entire surface of the NTCDA microcrystals was coated with Au nanoparticles approximately 20 nm in diameter (Fig. 4.17a) [16, 17].

4.4.3 *Revised Structural Trap Model*

As shown in Fig. 4.18b, a revised type of structural trap at the NTCDA/Au interface is illustrated by assuming that the Au particles were spherical, with the energy structure during multiplication (Fig. 4.18a). Considering that the NTCDA crystals were far larger (100 nm) than the Au particles (20 nm), noncontact sites and direct contact sites were inevitably formed between the NTCDA surface and Au spheres. The NTCDA surface offered blind alley sites that captured photogenerated holes under

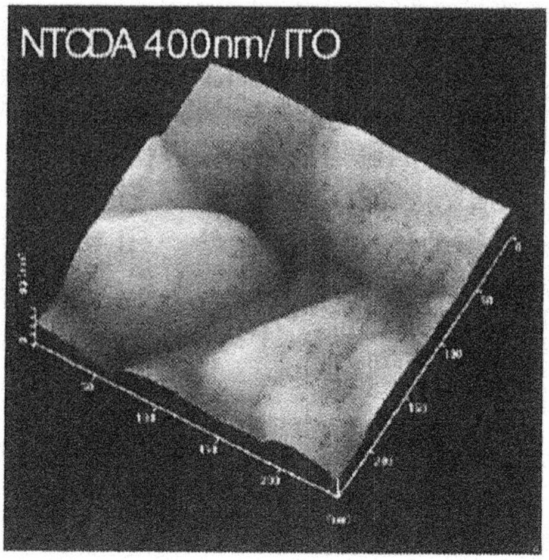

Fig. 4.16 Atomic force microscopy (AFM) image of 250 nm × 250 nm area for a 400 nm-thick NTCDA film deposited on the ITO glass substrate. The AFM images were obtained in contact mode in an ambient atmosphere. Reproduced with the permission of [15]. © 2002 IEICE

the concentration of the electric field toward the metal. Owing to the spherical shape of Au, direct contact sites and spatial gaps were formed at the Au/NTCDA interface. The direct contact sites acted as electron-injecting sites for the Au nanoparticles. The spatial gap offered a hole-accumulating area. The lateral spatial separation of the hole-accumulating sites and the electron-injecting sites explained why the injected electrons did not recombine with the trapped holes during multiplication.

4.4.4 Three-Dimensional Energy Structure

The three-dimensional energy structure for the revised structural trap (Fig. 4.19) was illustrated by electrostatic potential calculations based on Poisson's equation under the assumption that the diameter of the Au sphere was 10 nm [18]. At the contact site of the organic/metal interface, where there were no accumulated holes, the conduction band (CB) of the organic semiconductor gradually bent and directly connected with the metal. Figure 4.18a corresponds to this energy structure at the direct contact site of the organic/metal interface. In the hole accumulation area, the CB sharply bent because of the accumulated holes. In addition, a vacuum level existed between the organic surface and metal sphere. These two different energy structures were periodically repeated. The electron tunneling injecting sites at the direct organic/metal contacts and hole accumulating sites at the indirect organic/vacuum/metal contacts were laterally separated to the interface.

The numerical simulation of the structural trap of the type shown in Fig. 4.18b showed that the accumulated charges remaining at the noncontact organic film interface provided a sufficiently high field for tunneling the charge injection, leading to the multiplication process [18]. In addition, by assuming a low surface mobility,

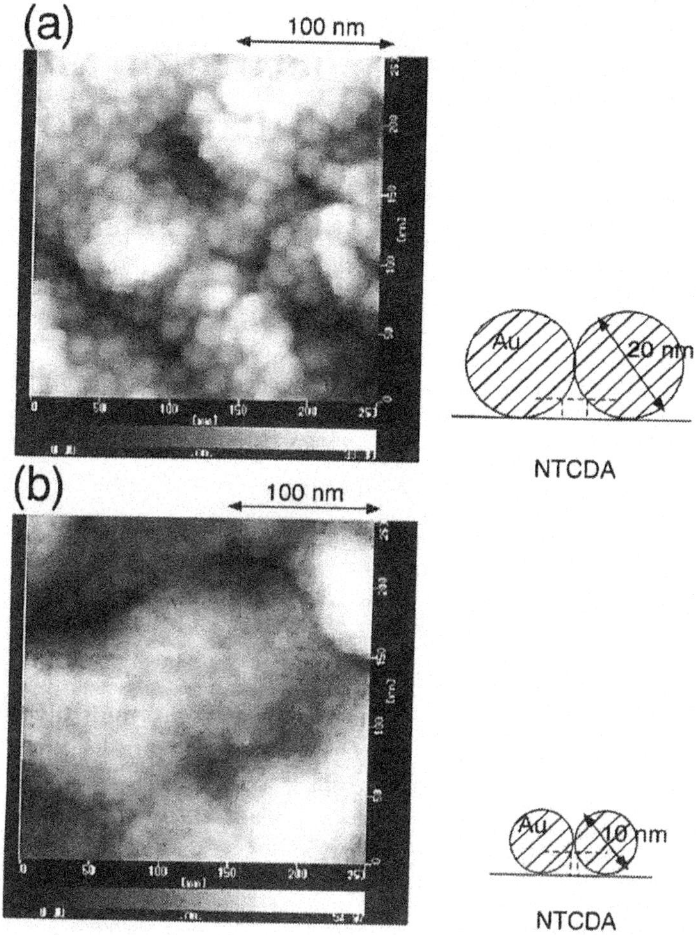

Fig. 4.17 AFM images of 250 nm × 250 nm area for the NTCDA films covered with 20 nm-thick Au films. Au was deposited from a resistivity-heated tungsten boat (**a**) and an EB evaporation source (**b**). Reproduced with the permission of [15]. © 2002 IEICE

compared with the bulk mobility, which was caused by the molecular-sized roughness at the organic surface (Fig. 4.18b), the large and slow response of the multiplied photocurrent (Chap. 3) was reproduced [18, 19].

4.4.5 Effects of Metal Nanoparticle Size

The revised structural trap model (Fig. 4.18b) suggested that the multiplication behaviors reflected the sensitivity to the size of the Au nanoparticles. Interestingly, when

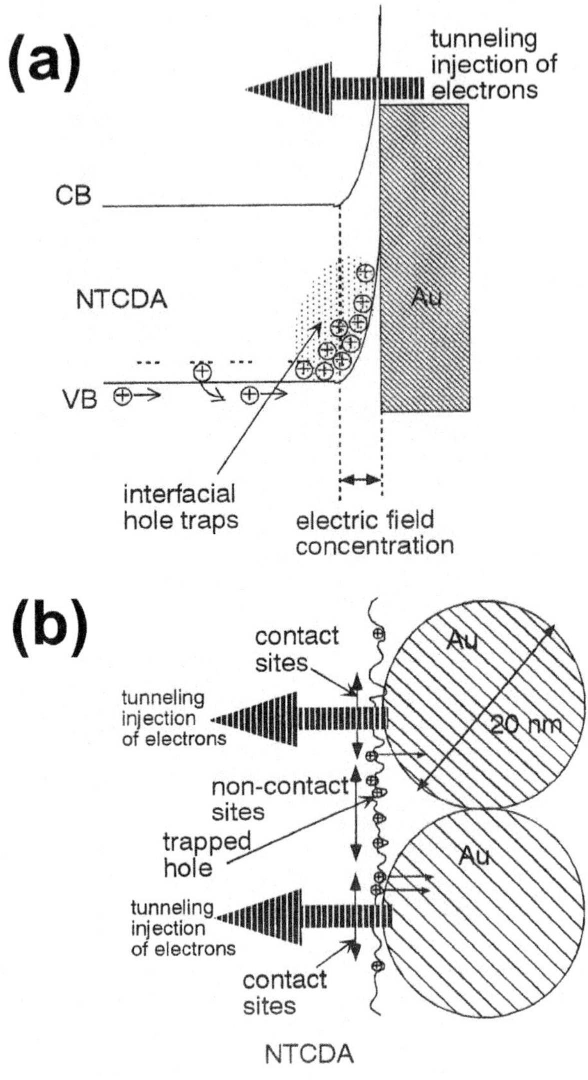

Fig. 4.18 Revised structural trap model. **a** Energy structure of the Me-PTC/Au interface during multiplication. **b** Schematic illustration of structural traps formed at the organic/metal interface. The holes were trapped by blind alleys when the electric field was toward the metal. Reproduced with the permission of [15]. © 2002 IEICE

Au was deposited from an electron beam (EB) evaporation source, multiplication was effectively suppressed (Fig. 4.15b, curve B). For the EB deposition (Fig. 4.17b), an Au particle diameter of approximately 10 nm was observed, which was considerably smaller than that of 20 nm deposited from the resistivity-heated source (Fig. 4.17a). Based on the aforementioned model, the multiplication suppression for EB deposition was due to a decrease in the gap area offering hole trap sites, as illustrated in Fig. 4.17a, b.

Another result was obtained by investigating the dependence of the multiplication behavior on the types of metal electrodes used in the ITO/NTCDA/metal cells [17,

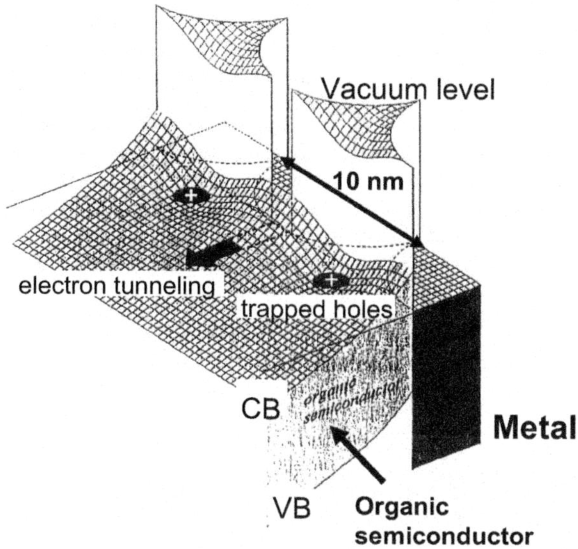

Fig. 4.19 Three-dimensional energy structure for the revised structural trap model (Fig. 4.18b). The diameter of the metal sphere was assumed to be 10 nm

20]. Pd, Ag, Au, Pt, Ti, and In were used as metals. They were deposited from an EB source. The multiplication rate was strongly dependent on the types of metals used. At 12 V, it reached 3000-fold for Pd and less than 1 for In (Fig. 4.20). Contrarily, the metal films, except In, comprised metal nanoparticles covering the entire NTCDA surface, as shown in Fig. 4.20b. The nanoparticle size depended on the type of metal used. The multiplication rate was observed to increase with the particle size (Fig. 4.20). The gap area that accumulated holes increased with the size of the metal nanoparticles. For In, no nanoparticles were observed by AFM. Therefore, it was concluded that In uniformly adhered to the entire surface of the NTCDA film and that there were no spatial gaps at the NTCDA/In interface, as shown in Fig. 4.20c.

Observed variety of NTCDA/metal interfaces is related to the various kinds of interactions between metals and organic films including NTCDA, which have been intensively studied [21–23].

A close relationship was observed between the multiplication behaviors and morphology of the metal films. The spherical shape of the deposited metal nanoparticles provided a spatial gap acting as a hole accumulation area, and direct metal/organic contacts acting as electron-injecting sites. This result strongly supported the revised structural trap model (Fig. 4.18b).

4.4.6 Rear Surface of Au and In Films

The existence of the spatial gap was confirmed by observing the rear surface of the Au film after peeling off the Au electrode from the NTCDA surface by using

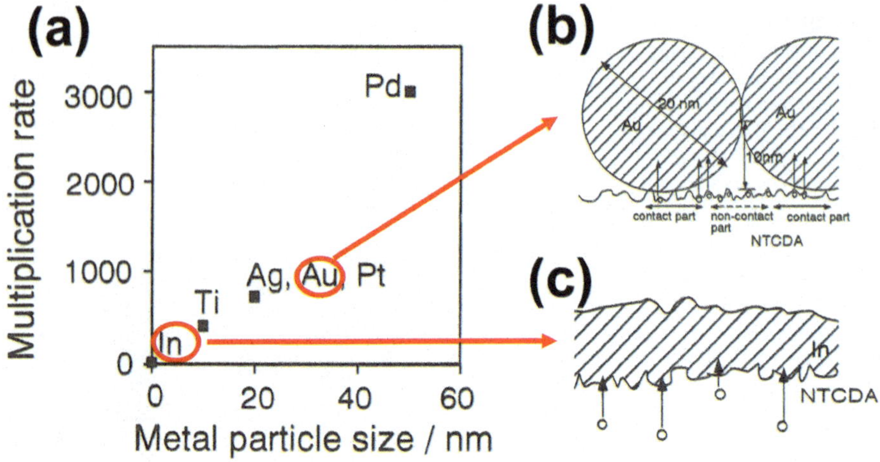

Fig. 4.20 **a** Dependence of the multiplication rate at 12 V on the metal nanoparticle size. The cell structure was the ITO/NTCDA/metals. The metals were Au, Pd, Ag, Pt, Ti, and In. The metal electrodes were negatively biased with respect to the ITO electrode. **b** Microscopic structures for the Au/NTCDA interface. **c** Microscopic structure of the In/NTCDA interface

adhesive tape [24]. Spherical structures were also observed for the rear surface of Au nanoparticles (Fig. 4.21a). This decisively proved the existence of spatial gaps at NTCDA/Au interface (Fig. 4.20b).

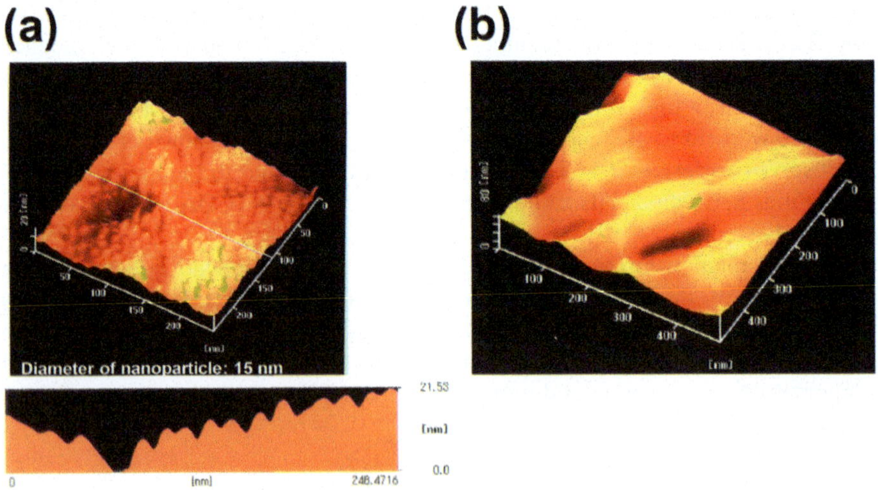

Fig. 4.21 **a** AFM image of the rear surface of the Au film after peeling off the Au electrode from the NTCDA surface. **b** AFM image of the rear surface of the In film after peeling off the In electrode from the NTCDA surface

On the contrary, the rear surface of the In film after peeling off the In electrode from the NTCDA surface showed no spherical structure and well coincides with the replica of NTCDA surface. This decisively proved that In uniformly adhered to the entire surface of the NTCDA film and that there were no spatial gaps at the NTCDA/In interface (Fig. 4.20c).

4.5 Essential Factors in the Structural Trap Model

The structural trap model (Figs. 4.1b and 4.18b) contained two essential factors for hole accumulation, i.e., the spatial gaps at the organic/metal interface and molecular-sized roughness at the organic film surface. For the organic/metal interface formed by the deposition of metal on the organic film, the former was related to the structure of the metal film, and the latter was related to the structure of the organic film.

4.5.1 Spatial Gap

As discussed in Sect. 4.4, the existence of spatial gaps at the organic/metal interface was confirmed. The spatial gaps were formed because of the spherical shape of the deposited metal nanoparticles. These spatial gaps were revealed to be indispensable for hole accumulation at the organic/metal interface, which triggered the photocurrent multiplication.

4.5.2 Molecular-Sized Roughness

As discussed in Sect. 4.2, the crystallization of the organic film was revealed to be indispensable for hole accumulation at the organic/metal interface. The surfaces of polycrystalline organic films that exhibit high photocurrent multiplication rates can be assumed to be microscopically rough. As shown in Fig. 4.18b, when the organic film surface exhibited molecular-sized roughness, the hole accumulation effectively occurred because the holes reaching the noncontact parts were captured by blind alleys, which acted as hole traps when the electric field was toward the metal (Fig. 4.8a) and prevented the lateral movement of holes. Contrarily, for the amorphous organic films showing minimal multiplication, hole accumulation hardly occurred because the holes reaching the noncontact parts easily moved along the microscopically smooth surface and escaped into metal particles (Fig. 4.8b).

4.5.3 *True Nature of Molecular-Sized Roughness*

The true nature of the molecular-sized roughness of the organic surface remains unanswered. Considering its close relationship with crystallinity, this is directly related to molecular stacking. The estimated values of 10 trapped holes per 10 nm × 10 nm area and blind alleys approximately 1 nm deep appeared to be related to the molecular-sized roughness. The following possibilities were considered: (i) The protrusion comprising a small number of molecules formed at the crystal surface. (ii) The roughness due to the stacked organic molecules themselves. (iii) The difference in carrier mobility in the direction perpendicular or parallel to the molecular plane in organic crystals formed by planar molecules, such as Me-PTC. For example, when the planar molecules were stacked face-to-face and the molecular plane was parallel to the crystal surface, the lateral movement of holes along the crystal surface, which was parallel to the molecular plane, appeared extremely difficult.

To resolve this, the molecular-level observation of the crystalline surface of the organic films was conducted.

This question will be answered in Chap. 5.

4.6 Conclusion

The revised structural trap model at the metal/organic interface has two essential factors for hole accumulation, which causes photocurrent multiplication. The first is the spatial gaps at the organic/metal interface. These spatial gaps were formed because of the spherical shape of the deposited metal nanoparticles. The second factor is the molecular-sized roughness of the organic film surface. The formation of molecular-sized roughness was observed to occur only when the organic films were crystalline. Both factors were confirmed to be indispensable for photocurrent multiplication.

Acknowledgements The author greatly appreciates S. Ooki of Dainichiseika Color and Chemicals Manufacturing Co. Ltd. for supplying many perylene pigment derivatives. The author also appreciates the funding of CREST (Core Research for Evolutionary Science and Technology) (Research subject: Control of molecular-level nanostructure at organic/metal interface and multiplication-type photosensor) from JST (Japan Science and Technology Agency).

References

1. Hiramoto, M., Imahigashi, T., Yokoyama, M.: Photocurrent multiplication in organic pigment films. Appl. Phys. Lett. **64**, 187–189 (1994)
2. Hiramoto, M., Kawase, S., Yokoyama, M.: Photoinduced hole injection multiplication in p-type quinacridone pigment films. Jpn. J. Appl. Phys. **35**, L349–L351 (1996)

3. Katsume, T., Hiramoto, M., Yokoyama, M.: Photocurrent multiplication in naphthalene tetracarboxylic anhydride film at room temperature. Appl. Phys. Lett. **69**, 3722–3724 (1996)
4. Hiramoto, M., Yokoyama, M.: Organic photocurrent multiplication device. Japanese Patent, No. 100797 (2002)
5. Nakayama, K., Hiramoto, M., Yokoyama, M.: Direct tracing of the photocurrent multiplication process in an organic pigment film. J. Appl. Phys. **84**, 6154–6156 (1998)
6. Nakayama, K., Hiramoto, M., Yokoyama, M.: Transient response of multiplied photocurrent observed in metal/organic pigment film interface. In: Proceedings of IS&T's NIP14: International Conference on Digital Printing Technologies, Toronto, Ontario, Canada, pp. 490–493, Oct. 18–23 (1998)
7. Katsume, T., Hiramoto, M., Yokoyama, M.: Detailed analysis of photocurrent multiplication phenomenon in organic pigment films. In: Proceedings of IS&T's NIP12: International Conference of Digital Printing Technologies, San Antonio, Texas, pp. 248–251, Oct. 27-Nov. 1 (1996)
8. Hiramoto, M., Nakayama, K., Katsume, T., Yokoyama, M.: Field-activated structural traps at organic pigment/metal interfaces causing photocurrent multiplication phenomena. Appl. Phys. Lett. **73**, 2627–2629 (1998)
9. Nakayama, K., Hiramoto, M., Yokoyama, M.: Photocurrent multiplication at organic/metal interface and surface morphology of organic films. J. Appl. Phys. **87**, 3365–3369 (2000)
10. Hiramoto, M., Katsume, T., Yokoyama, M.: Light amplification device based on organic multilayered films. Oyo Butsuri **64**, 1036–1039 (1995) (in Japanese)
11. Hiramoto, M., Nakayama, K., Sato, I., Kumaoka, H., Yokoyama, M.: Photocurrent multiplication phenomena at organic/metal and organic/organic Interfaces. Thin Solid Films **331**, 71–75 (1998)
12. Wagner, H.J., Loutfy, R.O., Hsiao, C.: J. Mater. Sci. **17**, 1982–2781
13. Gregg, B.A.: Evolution of photophysical and photovoltaic properties of perylene bis(phenethylimide) films upon solvent vapor annealing. J. Phys. Chem. **100**, 852–859 (1996)
14. Mizuguchi, J.: Electronic characterization of N, N'-bis(2-phenylethyl)perylene-3,4,9,10-bis(dicarboxyimide)and its application to optical disks. J. Appl. Phys. **84**, 4479–4486 (1998)
15. Hiramoto, M., Sato, I., Nakayama, K., Yokoyama, M.: Photocurrent multiplication at organic/metal interface and morphology of metal films. Jpn. J. Appl. Phys. **37**, L1184–L1186 (1998)
16. Hiramoto, M., Sato, I., Yokoyama, M.: Metal nanostructure of metal/organic interface causing photocurrent multiplication phenomenon. IEICE Trans. Electron. **E85-C**(6), 1253–1255 (2002)
17. Hiramoto, M.: Molecular-sized structural trap at organic-metal interface and photocurrent multiplication phenomenon. In: Salaneck, W.R., Seki, K., Kahn, A., Pireaux, J.-J. (eds.) Conjugated Polymer and Molecular Interfaces, Chap. 18, Marcel Dekker Inc., New York/Basel, pp. 585–612 (2002)
18. Nakayama, K., Hiramoto, M., Yokoyama, M.: Numerical simulations of photocurrent multiplication phenomenon at organic/metal interface. In: Proceedings of IS&T's NIP15: International Conference on Digital Printing Technologies, Orland, Florida, pp. 743–746, Oct. 17–22 (1999).
19. Nakayama, K., Hiramoto, M., Yokoyama, M.: Numerical simulations of photocurrent multiplication phenomenon process at organic/metal interface. In: Proceedings of Annual Conference of the Imaging Science of Japan, Tokyo, pp. 141–144, July 21–23 (1999) (in Japanese)
20. Nakayama, K., Niguma, Y., Hiramoto, M., Yokoyama, M.: Metal electrode dependence of photocurrent multiplication at organic/metal interface. In: Abstract of Research Center for Materials Science (RCMS) Workshop on Electronic Structures of Organic Interfaces (ESOI) 99, Nagoya, Japan, June 12–13, pp. 44–48 (1999)
21. Ishii, H., Sugiyama, K., Ito, E., Seki, K.: Energy level alignment and interfacial electronic structures at organic/metal and organic/organic interfaces. Adv. Mater. **11**, 605–625 (1999) and references therein

22. Cahen, D., Kahn, A., Umbach, E.: Energetics of molecular interfaces. Mater. Today **8**, 32–41 (2005)
23. Otero, R., Vázquez de Parga, A.L., Gallego, J.M.: Electronic, structural and chemical effects of charge-transfer at organic/inorganic interfaces. Surf. Sci. Rep. **72**, 105–145 (2017)
24. Yoshida, M., Nakayama, K., Hiramoto, M., Yokoyama, M.: Direct observation of the metal/organic interface causing photocurrent multiplication. In: Abstract of the 61th Autumn Meeting of Japan Society of Applied Physics, Tokyo, no. 3, p. 1075 (2000) (in Japanese)

Chapter 5
Photocurrent Multiplication in Organic Single Crystals—Molecular Blind Alleys

5.1 Background

A large 10^4-fold photocurrent multiplication in various organic semiconductor films was observed [1–4] (Chap. 2). A mechanism was suggested based on the tunnel injection of electrons from the metal electrode to the organic semiconductor film by accommodating trapped holes near the organic/metal interface (Fig. 5.7). In addition, in Chap. 3, a structural trap model was proposed (Fig. 5.8) [5], wherein the blind alleys at an imperfect organic/metal contact capture holes tightly under a strong electric field. Due to verified strong relationships between the morphology of metal/organic interfaces and multiplication behavior [6–9], a revised structural trap model with molecular-size roughness on the surface of crystalline organic films (Fig. 5.1) was provided in Chap. 4 [8].

5.2 Motivation

The origin of molecular-size roughness at the surface of crystalline organic semiconductor films has remained unknown [6, 8, 9]. However, interestingly, the previous results suggest that the multiplication was only observed in the crystalline films, and not in amorphous ones [6, 8, 9]. Accordingly, the author was intrigued by atomic scanning microscopy (AFM) results to explore the photocurrent multiplication in organic single-crystals [10–12]. The results of this study were expected to reveal the source of molecular-size roughness, which is estimated to be independent of complex structures (for instance, grain boundaries) and the packing of non-aligned small crystals.

M. Hiramoto, *Organic Photocurrent Multiplication*, Electronic Materials: Science & Technology, https://doi.org/10.1007/978-981-99-1237-7_5

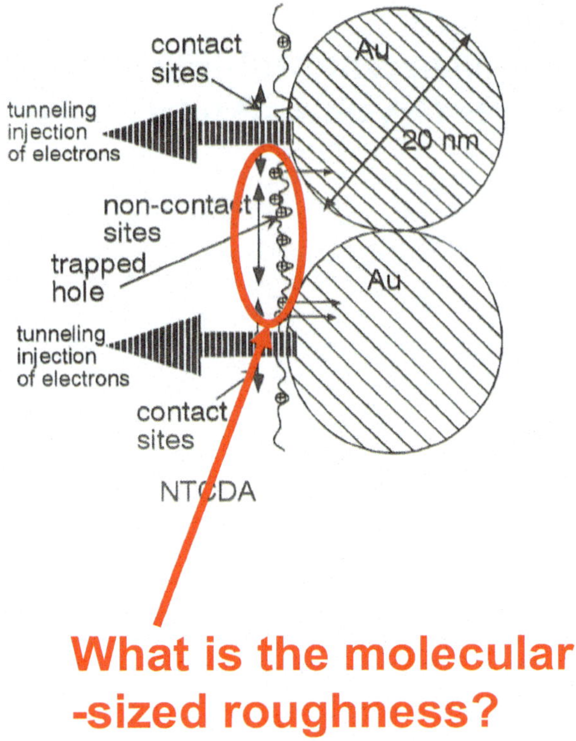

Fig. 5.1 Schematic illustration of revised structural trap model proposed in Chap. 4. Holes are trapped by the molecular-size roughness on the surface of crystalline organic semiconductor films. (However, the origin of the molecular-size roughness has not been clarified). Reproduced with the permission of [8]. © 2002 IEICE

5.3 Multiplication Behaviors of Organic Single Crystals

5.3.1 Naphthalene Tetracarboxylic Anhydride (NTCDA)

Except for an early report by Kallman and Pope in 1960 [13], studies on current multiplication in an organic crystal are scarce. NTCDA vacuum-deposited polycrystalline films are reported to demonstrate a large photocurrent multiplication of up to 10^5-fold and a substantial crystal size of approximately 100 nm [3, 6, 8]. Therefore, NTCDA was selected to explore the photocurrent multiplication phenomenon in organic single crystals (Figs. 5.2 and 5.4).

5.3.2 Single-Crystal Growth

NTCDA single crystals were grown according to the method described by Kloc et al. [14, 15] by using a sublimation apparatus in a temperature-gradient furnace to purify the organic semiconductors (Fig. 5.2a, b) [16]. To achieve high purity, sublimation

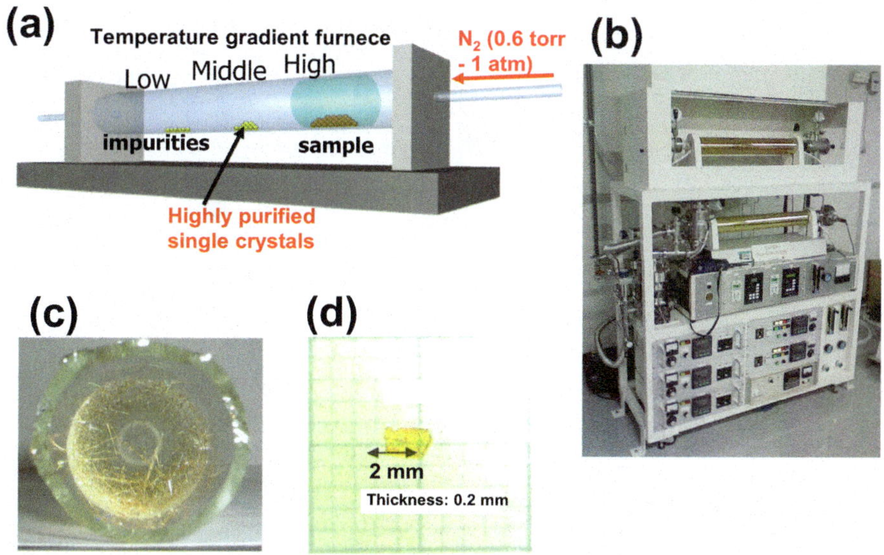

Fig. 5.2 **a** Schematic illustration of temperature gradient furnace. **b** A photograph of temperature gradient furnace. Two furnaces were installed. **c** A photograph of NTCDA single crystals grown inside the glass tube. **d** A photograph of NTCDA single crystal used in the sandwich-type cell that showed the first photocurrent multiplication

was performed under a high vacuum diffusion condition (10^{-3} Pa) and led to the formation of pure sample powder. For the single-crystal growth, the sublimation was performed under N_2 gas convection with a pressure range of 0.6–760 Torr. A glass tube was placed in the furnace. The NTCDA sample powder was positioned in the furnace at 330 °C temperature zone. Under these conditions, several NTCDA single crystals were grown inside the glass tube at 240 °C temperature zone (Fig. 5.2c) [17]. The typical dimensions of a single crystal were 2 mm × 1 mm, with a thickness of approximately 0.2 mm. Figure 5.2d presents an image of the NTCDA single crystal that demonstrated photocurrent multiplication behavior.

5.3.3 Cell Fabrication

Sandwich-type cells utilizing NTCDA single crystals (Fig. 5.4, inset) were fabricated using the procedure shown in Fig. 5.3. First, a single crystal was cut into two pieces (Fig. 5.3a). One was employed in cell fabrication, while another was used for AFM measurements. Because of the small size of single crystals and challenges in their handling, crystals were glued and surrounded by the epoxy resin on the polymer sheet with a smooth surface (Fig. 5.3b). After peeling off the sample, 20 nm-thick Au films were deposited by vacuum evaporation on both sides of the crystals (Fig. 5.3c). This

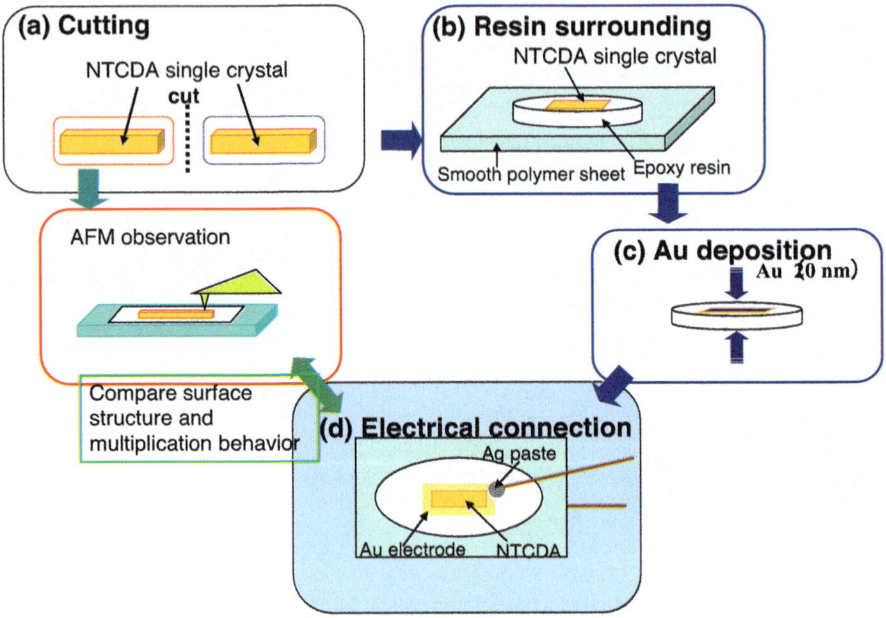

Fig. 5.3 Procedure of single-crystal cell fabrication. **a** A single crystal was cut into two pieces. One was used for surface observation by AFM. The other was incorporated in a cell. **b** The crystal was surrounded by epoxy resins on a smooth polymer sheet. **c** Au evaporation from both sides on the sample which was peeled off from the sheet. **d** Electrical connection to Cu wire by Ag paste. Studies on multiplication behavior and surface nanostructure by AFM were performed on the same crystal

preparation method allowed a direct electrical connection to the two Au electrodes using Ag paste (Fig. 5.3d). It is noteworthy that the obtained sandwich-type cell is fully symmetric, and the two Au/NTCDA crystal interfaces can be considered identical.

5.3.4 Multiplication Rate

Figure 5.4 presents the relationship between the multiplication rate and the applied electric field. The photocurrent was measured by applying a voltage of up to 700 V under an ambient atmosphere. The photocurrent quantum efficiency, representing the multiplication rate, was calculated from the ratio of the number of carriers flowing through the device to the number of photons absorbed by the crystal. The multiplication rate exceeded unity below 6×10^3 Vcm^{-1} and increased by 200-fold at 4×10^4 Vcm^{-1}. Although the strength of the electric field for the single-crystal cell (thickness = 167 μm), which was of the order of 10^4 Vcm^{-1}, was significantly smaller than that for the thin film (thickness = 0.5 μm), which was of the order of

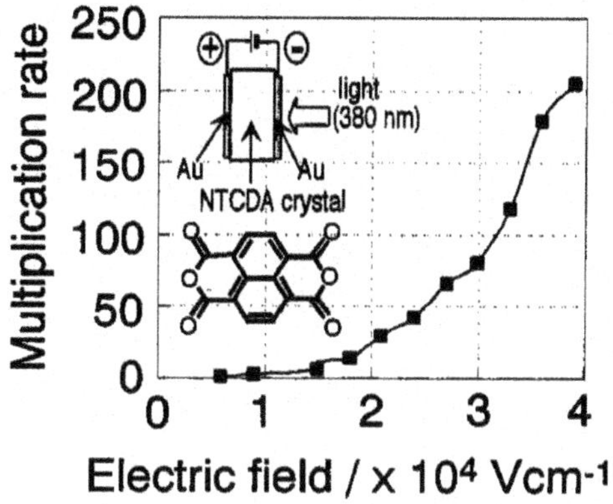

Fig. 5.4 Dependence of the multiplication rate on the applied electric field for an Au/NTCDA single crystal (167 μm)/Au cell. Monochromatic light of 380 nm (0.027 $mWcm^{-2}$) was irradiated on the Au electrode, which was negatively biased with respect to the counter Au electrode. The cell structure and chemical formula of NTCDA are shown in the inset. Reproduced with the permission of [17]. © 2002 American Institute of Physics

10^5 Vcm^{-1}, a relatively high multiplication rate could be observed. Higher electric fields could not be applied due to the significant thickness of single crystals.

5.3.5 *Spectral Sensitivity*

Figure 5.5a shows the spectral dependence of the multiplication rate. When the crystal was irradiated on the negatively biased Au electrode, the multiplication rate reached a maximum value of over 20-fold at 360 nm (Fig. 5.5a, curve A). The spectral region where the multiplication occurred was well aligned with the absorption spectrum of NTCDA (Fig. 5.5b). Interestingly, when the positively biased Au electrode was irradiated with light, the multiplication rate fell below unity throughout the spectrum (Fig. 5.5a, curve B). Since the 100 nm-thick NTCDA film has an absorbance exceeding unity in the peak region (Fig. 5.5b), a significant portion of the irradiated light would be absorbed within 100 nm (0.1 μm) of the Au electrode. Both the negatively (curves A) and positively (curves B) biased Au/NTCDA interfaces were selectively excited, assuming that only the irradiated side will be affected due to the thickness of the crystals (167 μm). Only light excitation on the crystal interface adjacent to the negatively biased electrode led to multiplication. Thus, it was concluded that the multiplication process takes place at the negatively biased NTCDA/Au interface. Because the device structure is symmetrical, the active Au/NTCDA interface for multiplication is determined only by the polarity of the bias. The results are in agreement with previously reported studies on NTCDA thin films sandwiched between Au and indium tin oxide (ITO) electrodes [3]. Therefore, it was established that the

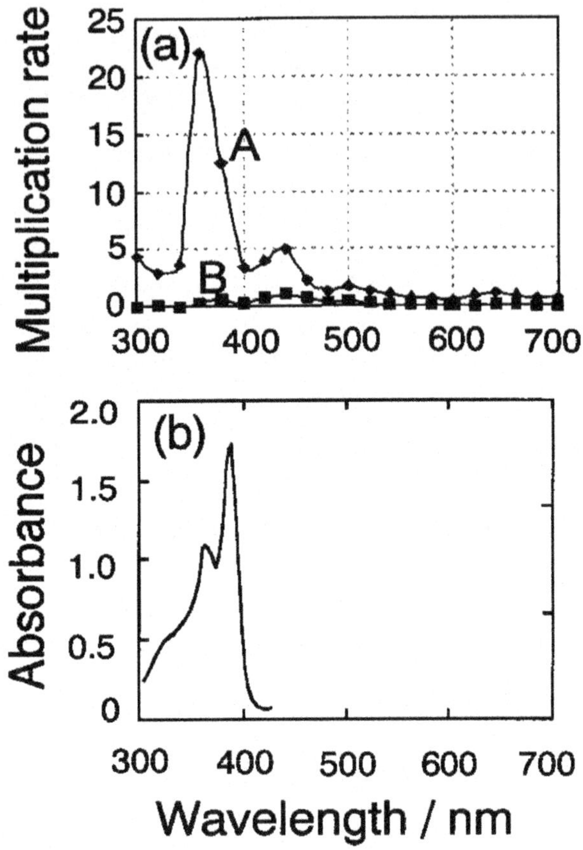

Fig. 5.5 **a** Spectral dependence of the multiplication rate for an Au/NTCDA single-crystal/Au cell using an applied field of 3.2×10^4 Vcm^{-1}. The light was irradiated on the negatively (curve A) and positively (curve B) biased Au. **b** The absorption spectrum of a 100 nm-thick NTCDA film. Reproduced with the permission of [17]. © 2002 American Institute of Physics

same photocurrent multiplication process (Fig. 5.7a) occurs at the single-crystal and thin-film NTCDA/Au interfaces.

5.3.6 Response of Multiplied Photocurrent

The most distinctive feature of single-crystal cells is their high-speed photoresponse (Fig. 5.6a). In this context, the response time is defined as the time required to reach 90% of the maximum value of the multiplied photocurrent. The response time from commencing the light irradiation was only 500 ms. In contrast, in thin-film devices fabricated by vacuum evaporation [3], the multiplied photocurrent did not reach a maximum, even after 60 s under the same electric field (Fig. 5.6b).

On the negatively biased Au electrode, the response time of the single-crystal device was found to be dependent on the irradiation wavelength. For monochromatic light at 380, 400, and 420 nm under an electric field of 3.2×10^4 Vcm^{-1}, the response

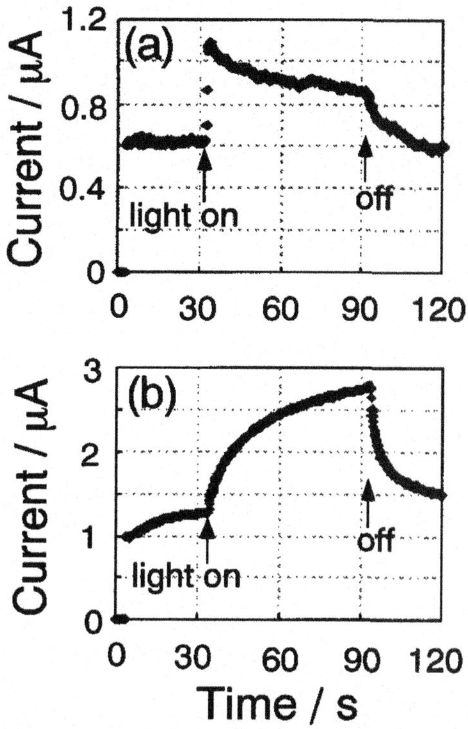

Fig. 5.6 Response profiles of the multiplied photocurrent observed for **a** a single-crystal cell (Fig. 5.4), and **b** for an ITO/NTCDA thin film (500 nm)/Au cell using the same measurement conditions: applied electric field = 3.2×10^4 Vcm^{-1}, light wavelength = 380 nm, light intensity = 0.027 $mWcm^{-2}$ (The light was irradiated on the negatively biased Au electrode. The multiplication rate was 120- and 50-fold, respectively). Reproduced with the permission of [17]. © 2002 American Institute of Physics

times were 0.5, 7, and 30 s, respectively. At 380, 400, and 420 nm, the light penetration depths for 90% of the photons absorbed by the NTCDA were 70, 220, and >1430 nm, respectively. The photoresponse was delayed with increasing penetration depth. As a result, photocarrier generation is favored in the region closest to the negatively biased interface to produce a higher-speed response.

5.3.7 Multiplication Mechanism

The photocurrent multiplication phenomenon in the NTCDA single crystals was discovered. The multiplication mechanism was concluded to be the same as that observed for vacuum-deposited NTCDA thin films (Fig. 5.7a). Multiplication occurs

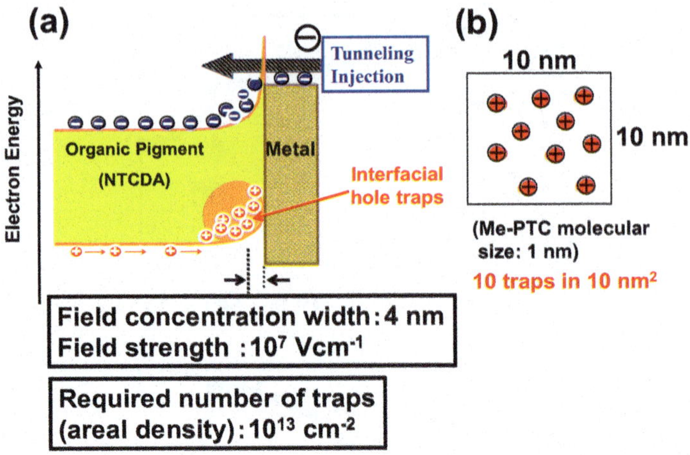

Fig. 5.7 **a** Energy structure of organic/metal interface during multiplication. **b** Anomalous areal density. Ten holes are trapped in the area of 10 nm × 10 nm

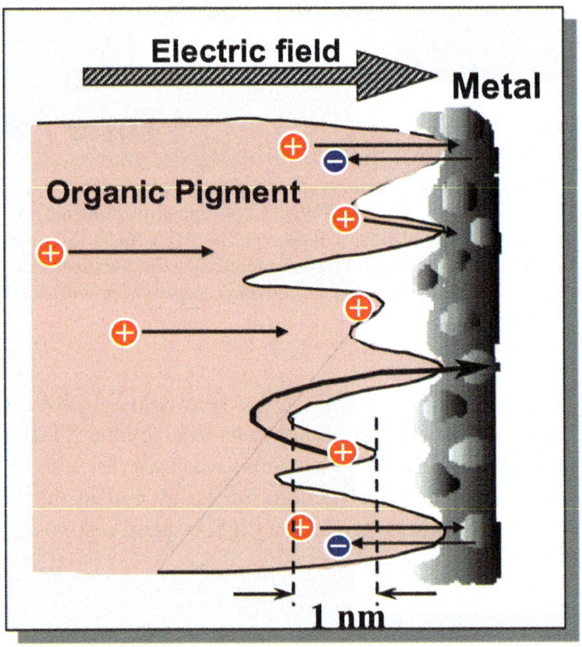

Fig. 5.8 Structural trap model at organic/metal interface based on hypothesis presented by the author. Reproduced with the permission of [5]. © 1998 American Institute of Physics

at the negatively biased single-crystal NTCDA/Au interface (Fig. 5.7a), where photogenerated holes move towards the negative electrode and are captured near the NTCDA/Au interface. The accumulation of trapped holes triggers the concentration of the electric field at the interface. Finally, tunneling-injection of electrons from the

Au electrode occurs, resulting in a significant multiplication. For thick single crystals, the absence of multiplication upon irradiation of the positively biased Au electrode (Fig. 5.5a, curve B) arises from the difficulty of hole delivery to the negatively biased interface through a distance of several hundred microns in the crystal bulk. The fast observed response time when irradiation occurs close to the active interface is attributed to the short travel distance for holes to reach the interface. Therefore, it can be concluded that, under the current operating conditions, the initial response time of the crystal multiplication devices is determined by the hole transport to the interface.

The reason for the higher speed response observed in the single crystals is not yet fully understood. One possible explanation is the lack of a grain boundary, which enables high hole mobility and results in a quick hole delivery to the interface. The second possible reason is related to the origin of the molecular-size roughness on the NTCDA single-crystal surface, which is also the central hypothesis of the present study (Fig. 5.1b). The hole accumulation process is expected to be completed rapidly for a single crystal.

5.4 Real Nature of Structural Trap

The research results on the structural trap causing multiplication are thoroughly discussed in Chaps. 3 and 4.

5.4.1 Anomalous Characters of Traps at Organic/Metal Interface

The leading cause of photocurrent multiplication is the carrier traps at the organic/metal interface. The detailed analyses in Chap. 3 provided two anomalous trap characters.

(i) **Energetic depth**: The energetic depth of traps, evaluated by the thermally stimulated current (TSC) measurements, increased and reached around 1 eV by increasing applied electric field towards the organic/metal interface (Chap. 3, Fig. 3.11b) [5]. This behavior is contrary to that of an ordinary electronic trap, for which the energetic depth declines by increasing applied electric field because of the Poole–Frenkel effect (Fig. 5.9a) [18–22].

(ii) **Areal density**: The areal density of traps reached 10^{13} cm^{-2}, which according to the tunneling-injection analysis by the Fowler–Nordheim (FN) equation, suggests that 10 holes are trapped in a 10 nm × 10 nm area (Chap. 3, Fig. 5.7b) [5, 23–26]. However, considering the molecular size of Me-PTC (1 nm), it is challenging to explain the extremely high areal density of traps formed by the impurities.

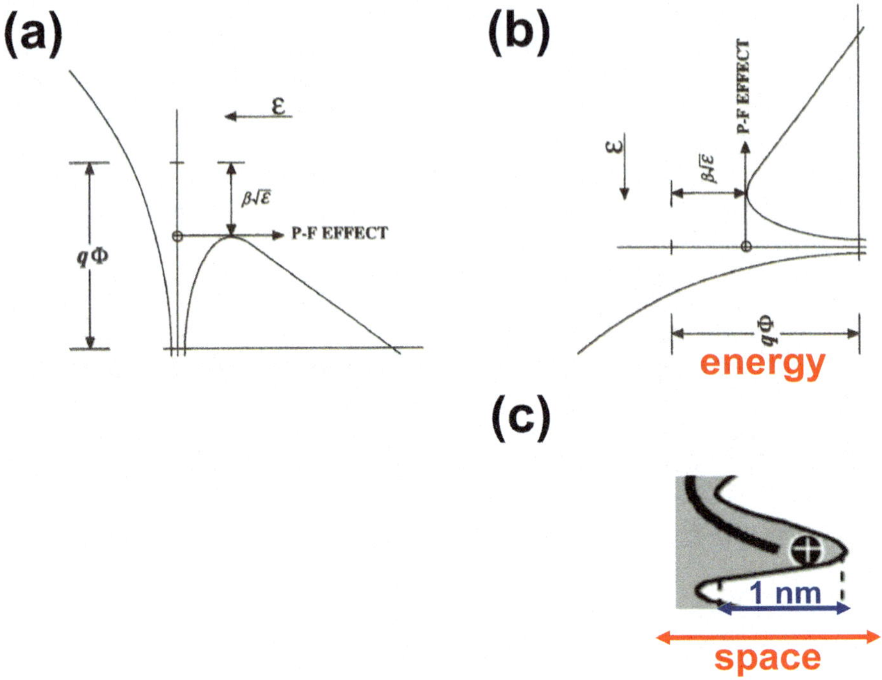

Fig. 5.9 **a** Energetic illustration of Poole–Frenkel effect according to the literature [21]. **b** 90° rotation of the illustration. **c** Replacement of energy with space for the rotated vertical axis. A dead-end is appeared. Reproduced with the permission of [21]. © 1999 Elsevier Science S. A

Therefore, the author concluded that ordinal electronic traps are not the main reason behind the multiplication phenomenon.

5.4.2 Structural Trap Model Hypothesis

Based on the results, a new mechanism was hypothesized (Chap. 3, Fig. 5.8), and a schematic illustration of this mechanism is presented in Fig. 5.8. The surface of organic films with high roughness is unable to create uniform junctions with metals, leading to several molecular-size blind alleys (dead ends) at the organic/metal interface. As a result, photogenerated holes can accumulate in these areas. Assuming a 1 nm depth for these blind alleys and the concentrated electric field of 10^7 Vcm^{-1}, the energetic depth of the trap reached 1 eV—the energy required for the escape of a hole from the blind alley against the electric field. In agreement with TSC measurements, the energetic depth of the trap increases by intensifying the electric field in this model.

5.4.3 The Idea Behind the Hypothesis

In reality, once holes are trapped in the TSC experiment, they strongly resist escaping even under a very intense collecting voltage. However, in the Poole–Frenkel type traps [18–22], the holes can readily escape because the energetic trap depth declines under a stronger electric field (Fig. 5.9a). Due to this discrepancy, I could not help but wonder about the existence of a particular type of trap that is different from ordinary traps with a simple energetic level in the bandgap such as impurities. My research experiences warned me that this substantial discrepancy implies a new concept. Suddenly, I imagined a 90° rotation of the Poole–Frenkel effect energetic illustration (Fig. 5.9b) [21] and the replacement of the rotated vertical axis from energy to space (Fig. 5.9b, c). A little later, I realized that the closed ends could serve as a trap under the applied electric field. I was rushed to calculate the energy required to escape from such new trap types by assuming a molecular size of 1 nm (10^{-7} cm) and a concentrated electric field of 10^7 Vcm^{-1} (Fig. 5.9c). The obtained value of 1 eV was close to the observed energetic depth of the TSC measurements under the highest applied voltage. Based on these considerations, I envisioned an imperfect junction at the organic/metal interface (Fig. 5.8). The author has the impression that geometrical thinking can often create original ideas.

5.4.4 Physically Existing Nanostructure Acting as Structural Trap

Two distinct characteristics of the structural trap at the organic/metal interface (Fig. 5.8) are as follows:

(i) Spatial gap.
(ii) Blind alleys with 1 nm spatial depth.

Identification of existing nanostructures, spatial gaps, and blind alleys requires persistence.

5.4.5 Spatial Gap

Physically existing nanostructures at the organic/metal interface were identified as the spatial gaps caused by the spherical shape of the metal nanoparticles (Chap. 4) [6–9]. These nanostructures can form for both organic films and organic single crystals. The AFM image of a vacuum-evaporated Au film (20 nm) on the NTCDA single crystal is shown in Fig. 5.10b. The deposited film is identified as a pack of Au nanoparticles with a 20 nm diameter. The image shows spatial gaps between the organic surface

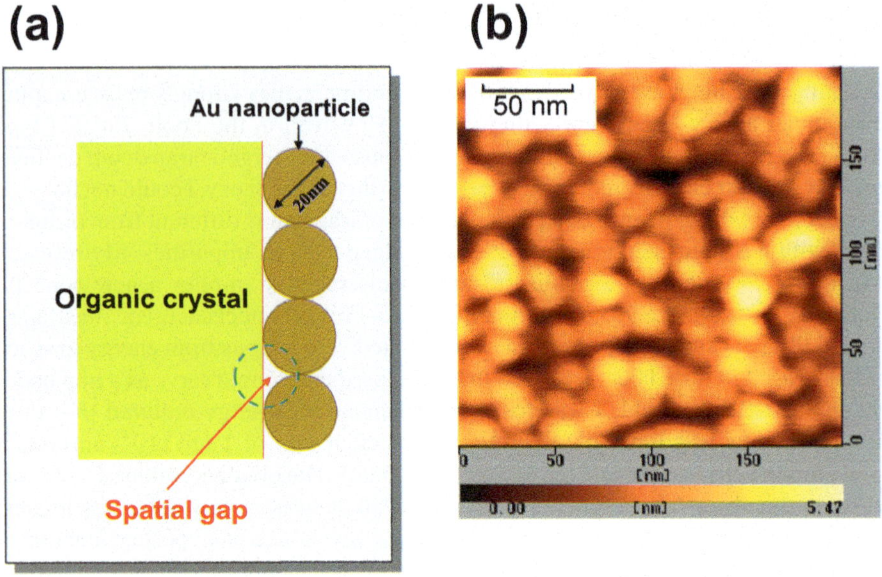

Fig. 5.10 **a** Schematic illustration of the NTCDA single-crystal/Au interface. Au nanoparticles with a diameter of 20 nm show a spherical shape. There are several spatial gaps between NTCDA and Au spheres. **b** AFM image of Au film (20 nm) deposited on the NTCDA single-crystal surface

and metal nanoparticles (Fig. 5.10a). Spatial gaps have been proven to be essential for photocurrent multiplication (Chap. 4) [6–9].

5.4.6 Molecular-Size Roughness

Photocurrent multiplication has only been observed for crystalline films, not the amorphous ones (Fig. 5.11) (Chap. 4) [6, 9]. This observation indicates the presence of molecular-size roughness (Fig. 5.1) on the organic crystal surface. This molecular-size roughness triggers the multiplication process and is one of the critical factors for the multiplication process (Chap. 4) [6, 9].

5.5 Molecular Blind Alley Model

5.5.1 Why Single Crystal?

Elucidating the origin of blind alleys for vacuum-deposited organic films is challenging [6–9]. However, it is interesting that multiplication is only observed for the

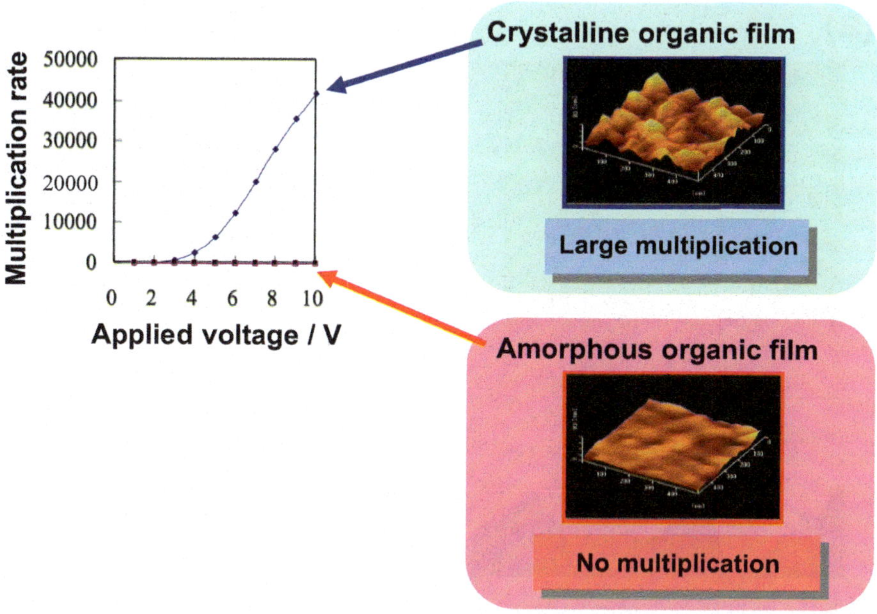

Fig. 5.11 Dependence of multiplication rate on applied voltage for crystalline (NTCDA) and amorphous (PhEt-PTC, Figs. 4.3 (curve c) and 4.4c in Chap. 4) organic films. Corresponding AFM images are also shown

crystalline films. Thus, the surface of organic single crystals showing multiplication was studied as a single crystal, with a molecularly flat surface and no grain boundaries to potentially expose the existing nanostructure by AFM measurements.

5.5.2 Molecular Steps

Figure 5.12a presents an AFM image of the NTCDA single-crystal surface, which has shown a multiplication rate of 100-fold at 2×10^4 Vcm^{-1} (Fig. 5.14, green curve). Several steps were detected in a 2 μm × 2 μm area. The step height of 0.5 nm (Fig. 5.12a) corresponds well with the molecular width of NTCDA (0.5 nm) (Fig. 5.12b). This observation is consistent with the formation of π-stacking NTCDA molecules parallel to the NTCDA crystal surface. Other molecular-size structures were not observed in the macroscopic image of the optical resolution to the nanoscopic image of the stacked molecules (Fig. 5.15, lower AFM image). Based on these observations, two possibilities for the origin of molecular-size roughness can be envisioned.

(i) Molecular step.
(ii) The roughness of NTCDA molecular stacking.

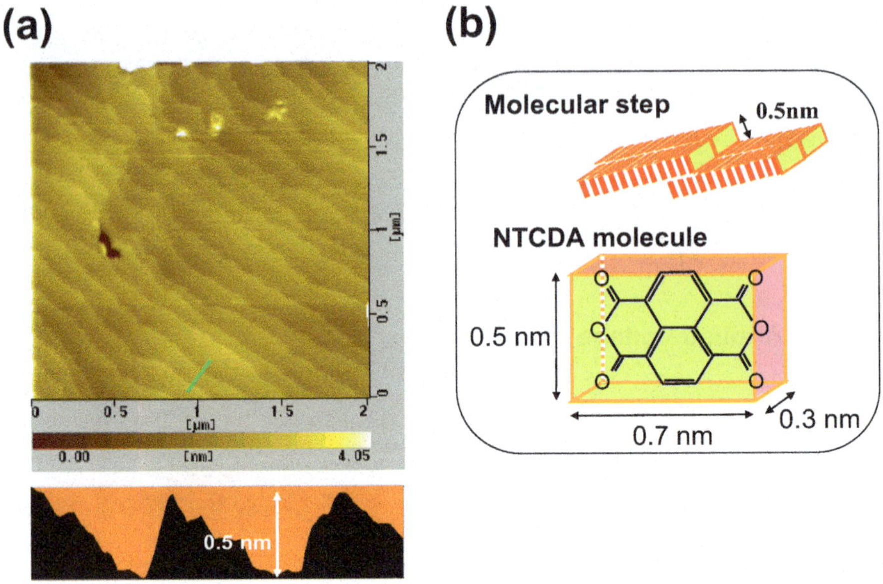

Fig. 5.12 **a** AFM image of the NTCDA single-crystal surface (2 μm × 2 μm). The cross-section of the green line in this image is also shown. **b** Schematic illustration of molecular step and an NTCDA molecule

According to the results in Sects. 5.4.3 and 5.4.4, it was determined that the former is the nanostructure that causes multiplication.

5.5.3 Molecular Steps Acting as Molecular Blind Alleys

The height of the detected molecular steps was 0.5 nm. The spatial depth of a blind alley of 1 nm in the structural trap model (Fig. 5.8) was calculated based on the concentrated electric field of 10^7 Vcm^{-1} evaluated by FN analysis and the energetic depth of trap of 1 eV evaluated by TSC measurements. The molecular step height (0.5 nm) appears to be related to the expected spatial depth of the blind alley (1 nm). Therefore, it was undeniably concluded that the molecular steps serve as blind alleys.

In Fig. 5.13, the nanostructure of the NTCDA/Au interface is illustrated on a real scale. The term "molecular blind alley" is coined for the molecular steps in this context. First, there are several spatial gaps behind the 20 nm spherical Au nanoparticles. Second, molecular steps with a spatial depth of 0.5 nm act as the molecular blind alleys.

Photogenerated holes reach the interface and are captured by molecular steps serving as molecular blind alleys. A strong electric field is concentrated at the interface during the multiplication process. In order to escape from the molecular blind alley, holes should experience the concentrated electric field of 10^7 Vcm^{-1}

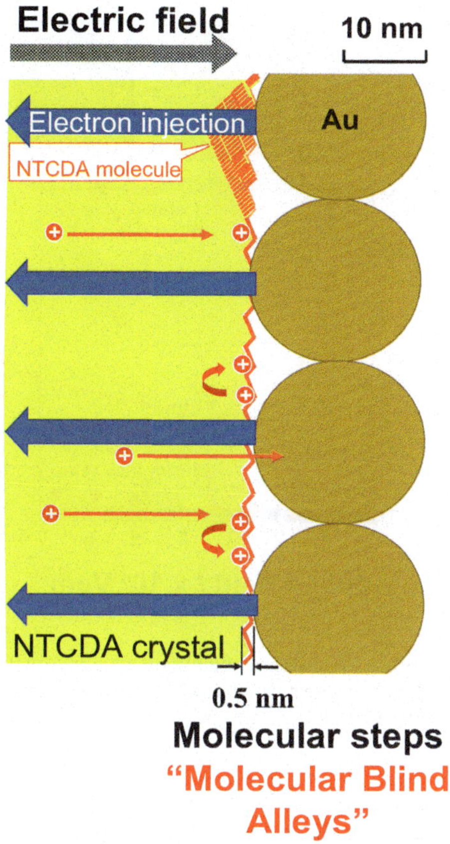

Fig. 5.13 Molecular blind alley model. The nanostructure of the NTCDA/Au interface is illustrated on a real scale

(Fig. 5.13, curved red arrows). As a result, the delivery of holes to the Au electrode is suppressed, and they are accumulated at molecular blind alleys under the spatial gaps. Finally, a high quantity of electrons is injected by tunneling from the contact sites of the Au nanoparticles (Fig. 5.13, blue arrows). This mechanism ultimately leads to photocurrent multiplication. Because there are no molecular steps on the surface of the amorphous layers, multiplication only occurs in the crystalline films.

5.5.4 Molecular Step Density Versus Multiplication Rate

The density of molecular steps on the NTCDA single-crystal surface can vary depending on crystal growth conditions. Figure 5.14 reveals the relationship between the multiplication rate and the applied electric field. AFM images of the crystal surface of cells are also presented. For example, the crystal surface grown under a low N_2 pressure (0.6 Torr) demonstrated the highest density of steps and also

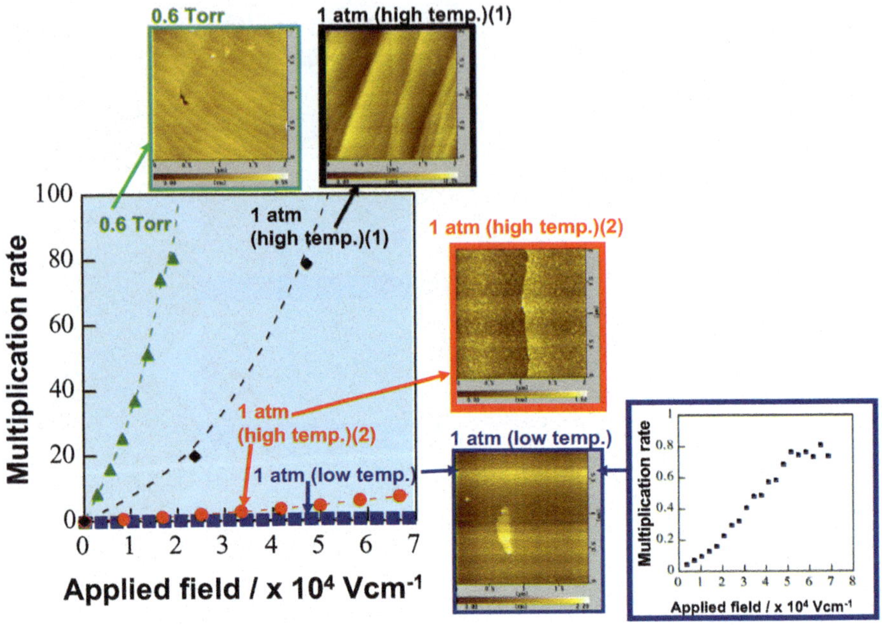

Fig. 5.14 Dependence of multiplication rate on the applied electric field for cells bearing NTCDA single crystals with various surface densities of molecular steps. AFM images for the crystal surface used for each cell are also presented. An expanded view for the blue curve is also shown

exhibited the highest multiplication rate (100-fold at 2×10^4 Vcm^{-1}) (Fig. 5.14, green). On the contrary, the crystal surface grown in the low-temperature region of the furnace and under a high N_2 pressure (760 Torr) had only a few steps and showed the lowest multiplication rate (0.8 at 5×10^4 Vcm^{-1}) (Fig. 5.14, blue). In addition, a moderate multiplication rate was observed for crystals with a medium density of steps (Fig. 5.14, black and red). Results suggest that a significant multiplication rate can be obtained by increasing the number of molecular steps. In addition, one single crystal bearing a minimal number of molecular steps showed no multiplication. These observations provide strong evidence for the importance of molecular blind alleys (molecular steps) in photocurrent multiplication.

5.5.5 *Molecular Kinks Acting as Molecular Blind Alley Sites*

To perform quantitative analysis, molecular stacking at the terrace part of the same single crystal was studied by AFM. The terrace part of the upper AFM image (Fig. 5.15a) was zoomed into the lower AFM image (Fig. 5.15b) without rotating the sample. Molecular stacking of the NTCDA having the molecular length of 0.7 nm (Fig. 5.12b, white lines) and the space between two NTCDA molecules, i.e., the

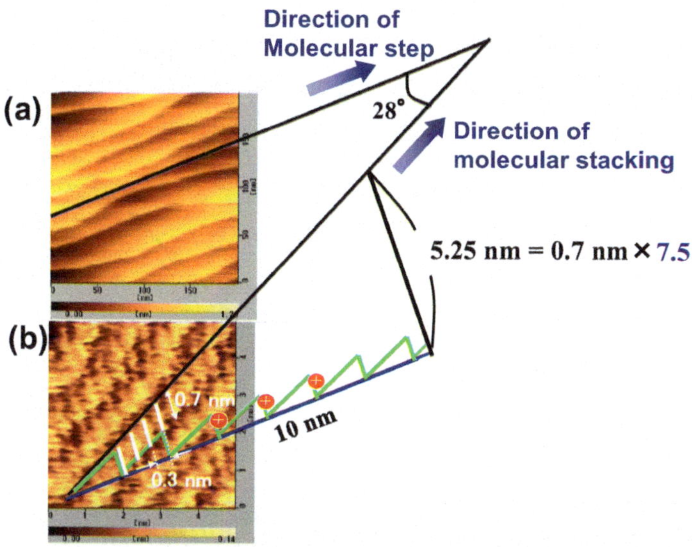

Fig. 5.15 **a** AFM image of the NTCDA single-crystal surface (2 μm × 2 μm). **b** AFM image of the molecular stacking at the terrace part for the same single crystal (5 nm × 5 nm) without rotation. The molecular length of 0.7 nm (Fig. 5.12b) is shown by white lines. The direction of molecular step and molecular stacking are out of alignment at 28°. Molecular kinks are schematically illustrated by the green line. Holes are trapped at kink sites

space between two white lines (0.3 nm) (Fig. 5.12b) was observed. The direction of the molecular step and that of molecular stacking were out of alignment at 28° (Fig. 5.15). To explain this direction difference, atleast 7.5 molecular kinks per 10 mm step length need to exist. Therefore, molecular kinks were inevitably introduced. The molecular kinks are schematically illustrated by the green line in Fig. 5.15. If the step lines were winding, the number of kinks would increase.

We assumed that the molecular kinks in the molecular step lines act as molecular blind alley sites because holes seem to be easily captured by these jags. Thus, the minimum number of molecular kinks (molecular blind alley sites) per unit area for the crystal surface in Fig. 5.15 could be calculated as 4.5×10^{11} cm^{-2}. The dependence of the multiplication rate at 2×10^4 Vcm^{-1} on the number of molecular blind alley sites is shown in Fig. 5.16. The multiplication rate increased with the number of molecular blind alley sites (molecular kinks). Therefore, it is reasonable that the molecular kinks in the molecular step lines act as molecular blind alley sites during the multiplication process.

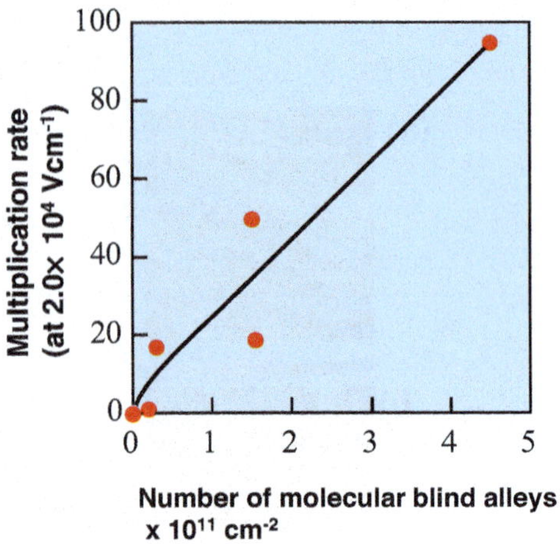

Fig. 5.16 Dependence of the multiplication rate at 2×10^4 Vcm^{-1} on the number of molecular blind alleys per 1 cm^2

5.5.6 Deliberate Formation of Molecular Blind Alleys

Molecular steps and kinks can be deliberately formed on the NTCDA crystal surface. An NTCDA crystal with minimal molecular steps showed very poor multiplication (Fig. 5.17a), blue curve a, and the AFM image). However, by deposition of 1 nm-thick NTCDA molecules on the same NTCDA crystal surface, many island structures with a height of 1 nm were formed on the surface (Fig. 5.17b), AFM image). As a result of this deposition, photocurrent multiplication was increased by 2,000-fold (Fig. 5.17b, red curve b). This result can be reasonably explained by the formation of molecular steps with kinks at the edges of islands, which can serve as molecular blind alleys. This experiment clearly demonstrates that multiplication behavior can be controlled by the deliberate formation of molecular blind alleys.

5.5.7 Molecular Blind Alleys in Polycrystalline NTCDA Films

Polycrystalline organic films, such as NTCDA and Me-PTC, which exhibit large photocurrent multiplication (Chaps. 2, 3, and 4), are composed of small crystal grains (Fig. 5.18) [1–9]. Each crystal grain has a high density of molecular steps and kinks (molecular blind alleys) on its curved surface (Fig. 5.18). Combined with the spatial gaps formed by spherical metal nanoparticles, these molecular blind alleys lead to a large photocurrent multiplication (Fig. 5.13).

It is noteworthy that the surface structure of small crystals in polycrystalline organic films (Fig. 5.18) is more complex than that of the organic single crystal. Obviously, there are several grain boundaries. Small crystals have random orientations and are not aligned along a particular direction. Several crystal planes with various densities of molecular steps and kinks are exposed on a crystal film surface. Compared

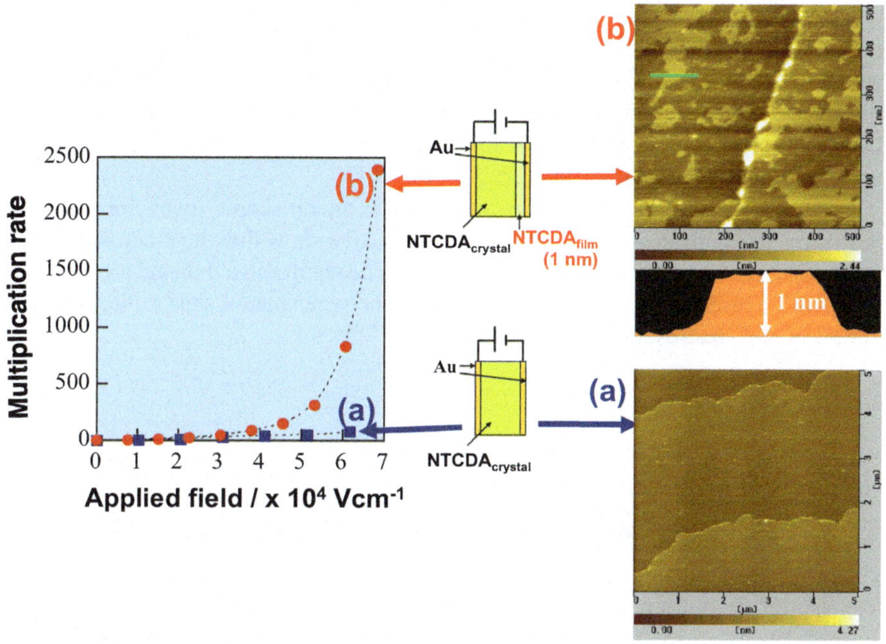

Fig. 5.17 Dependence of the multiplication rate on the applied electric field for the sandwich cells and AFM images of NTCDA single crystals. **a** A single crystal with minimal steps. **b** A single crystal bearing several islands prepared by the vacuum-deposition of 1 nm-thick NTCDA

Fig. 5.18 AFM image of vacuum-deposited NTCDA film consisting of several small crystals. Each crystal grain, which exposes various crystal planes (blue lines), has high density of the molecular blind alleys (molecular steps kinks) on its surface

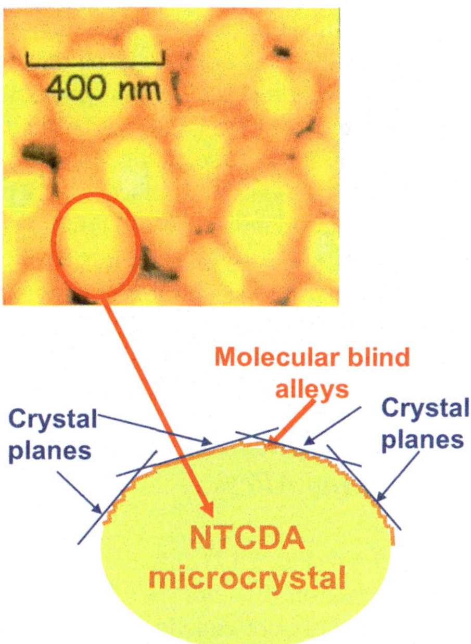

to organic single crystals, crystalline films bear a significantly larger number of molecular blind alleys. As a result, a remarkably large photocurrent multiplication (10^5-fold) can be observed for the organic crystalline films. The photocurrent multiplication phenomenon is expected to be more feasible for organic semiconductors than the inorganic counterparts because the size of organic molecules is typically larger than that of inorganic atoms.

Molecular-level nanostructures can lead to macroscopic behavior, such as photocurrent multiplication. Surprisingly, our results show that the macroscopic and nanoscopic phenomena are directly connected. The results also indicate a vast area of uncultivated research on the direct relationships between nanoscopic molecular-level structures and macroscopic device functions.

5.6 Future Perspective

5.6.1 *High Resolved AFM Observation of Steps and Kinks*

The spatial resolution of scanning probe microscopy (SPM) images of steps (Figs. 5.12 and 5.15) requires further improvements. By obtaining clear images of the molecular steps and kinks and taking into account the winding of steps, a precise number of molecular blind alleys can be measured. Moreover, a more detailed nanostructure of molecular blind alleys can be obtained.

In Fig. 5.19, two examples of explicit and high spatially resolved SPM images are presented. Figure 5.19a shows an AFM image of a pentacene single crystal [11], revealing that the detailed nanostructure of the resolved molecular steps is not simple. It is winding, and several prickles and the peninsula can be observed. The molecular-level kink structures can be distinguished by zooming on the step structure. Figure 5.19b displays a scanning tunneling microscopy (STM) image of a monolayer NTCDA on the Ag(110) surface [27]. The images of the NTCDA molecules are well resolved, and molecular level images of the steps and kinks are clearly realized. Similar precise observations are required for the NTCDA single crystals investigated in this study to further clarify the relationship between molecular blind alleys and the multiplication behavior.

5.6.2 *Multiplication Control via the Design of Molecular Blind Alleys*

Section 5.4.6 (Fig. 5.17) only provides preliminary results on the deliberate control of multiplication behavior by forming molecular blind alleys. As the height of the island (1 nm) is not exactly equal to the molecular dimensions (Fig. 5.12b), the observed islands are not homoepitaxially grown single crystals. Quite recently, the

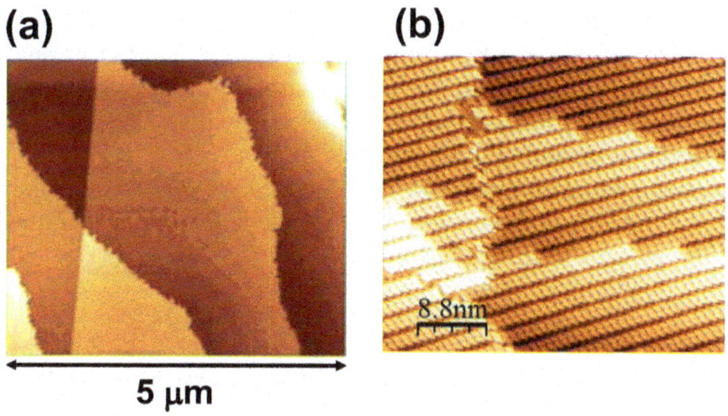

Fig. 5.19 **a** AFM image of a pentacene single crystal [11]. Detailed nanostructure of molecular steps is resolved. **b** STM image of monolayer NTCDA on Ag(110) surface [29]. Molecular-level observation of steps and kinks are achieved. Reproduced with the permission of [11] and [27]. © 2018 American Chemical Society

homoepitaxial growth of rubrene films on rubrene single crystals has been achieved by a prolonged deposition rate of 0.03 nm s^{-1} (Fig. 5.20) [12, 28, 29]. Alignment of hexagonal islands in the same direction and a step height of 1.3 nm, which coincides with the molecular height of rubrene, confirmed the homoepitaxial growth.

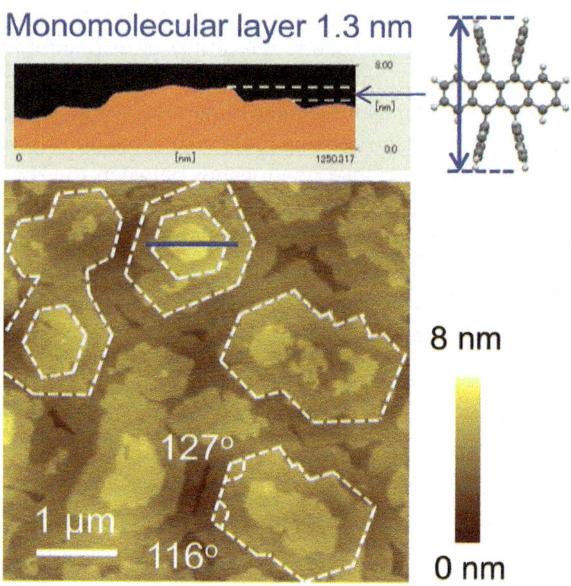

Fig. 5.20 Homoepitaxial growth of rubrene films on the rubrene single crystal by a very slow deposition rate of 0.03 nms^{-1} [12, 28, 29]. All hexagonal islands align along the same direction, and the step height of 1.3 nm coincides with molecular height of rubrene. Reproduced with the permission of [12]. © 2017, WILEY-VCH Verlag GmbH &CO. KGaA, Weinheim

Since the same homoepitaxial growth of NTCDA on the NTCDA single crystals is attainable, precise control of the number of molecular steps and kinks can be expected in future studies. Such precise control will lead to an improved multiplication behavior. The spatial depth of a molecular blind alley can be controlled at the nanoscale by changing the number of molecular monolayers. In addition, a molecular blind alley with extended spatial depth can be formed by utilizing organic molecules with large molecular sizes. All in all, deliberate control of multiplication by careful design of molecular blind alleys is feasible.

5.6.3 Design of Molecular Blind Alleys by Nanoimprint Lithography

Nanostructures with similar sizes to a molecular blind alley can be fabricated by nanoimprint lithography [30–33]. By exploiting nanoimprinting, molecular-size blind alleys with various shapes, sizes, and densities can be designed for inorganic semiconductors, amorphous organic films, and polymers. In addition, the design of molecular-size blind alleys can control the different aspects of multiplication behavior, such as the response rate.

5.7 Conclusion

We observed the photocurrent multiplication phenomenon in NTCDA single crystals. Photocurrent multiplication was discovered to be originated from the molecular steps and kinks acting as molecular blind alleys. The proposed structural trap model was revised to adopt the existing molecular blind alley hypothesis. The model was proven by discovering a close relationship between the molecular blind alleys and multiplication rate. The deliberate design of molecular blind alleys can control photocurrent multiplication behavior. Molecular-level nanostructures can lead to macroscopic functions, such as photocurrent multiplication. There is a huge area of uncultivated research on the nanoscopic molecular-level structures directly connecting to the macroscopic device functions.

References

1. Hiramoto, M., Imahigashi, T., Yokoyama, M.: Photocurrent multiplication in organic pigment films. Appl. Phys. Lett. **64**, 187–189 (1994)
2. Hiramoto, M., Kawase, S., Yokoyama, M.: Photoinduced hole injection multiplication in p-type quinacridone pigment films. Jpn. J. Appl. Phys. **35**, L349–L351 (1996)
3. Katsume, T., Hiramoto, M., Yokoyama, M.: Photocurrent multiplication in naphthalene tetracarboxylic anhydride film at room temperature. Appl. Phys. Lett. **69**, 3722–3724 (1996)

4. Hiramoto, M., Yokoyama, M.: Organic photocurrent multiplication device. Japanese Patent, No. 100797 (2002)
5. Hiramoto, M., Nakayama, K., Katsume, T., Yokoyama, M.: Field-activated structural traps at organic pigment/metal interfaces causing photocurrent multiplication phenomena. Appl. Phys. Lett. **73**, 2627–2629 (1998)
6. Hiramoto, M.: Molecular-sized structural trap at organic-metal interface and photocurrent multiplication phenomenon. In: Salaneck, W.R., Seki, K., Kahn, A., Pireaux, J.-J. (eds.) Conjugated Polymer and Molecular Interfaces (Chap. 18), pp. 585–612. Marcel Dekker Inc., New York/Basel (2002), and references therein
7. Hiramoto, M., Sato, I., Nakayama, K., Yokoyama, M.: Photocurrent multiplication at organic/metal interface and morphology of metal films. Jpn. J. Appl. Phys. **37**, L1184–L1186 (1998)
8. Hiramoto, M., Sato, I., Yokoyama, M.: Metal nanostructure of metal/organic interface causing photocurrent multiplication phenomenon. IEICE Trans. Electron. E85-C (6), 1253–1255 (2002)
9. Nakayama, K., Hiramoto, M., Yokoyama, M.: Photocurrent multiplication at organic/metal interface and surface morphology of organic films. J. Appl. Phys. **87**, 3365–3369 (2000)
10. Minato, T., Aoki, H., Fukidome, H., Wagner, T., Itaya, K.: High-resolution molecular images of rubrene single-crystals obtained by frequency modulation atomic force microscopy. Appl. Phys. Lett. **093302**, 3 pages (2009)
11. Sato, K., Sawaguchi, T., Sakata, M., Itaya, K.: Noncontact atomic force microscopy of perfect single-crystals of pentacene prepared by crystallization from solution. Langmuir **23**, 12788–12790 (2007)
12. Ohashi, C., Izawa, S., Shinmura, Y., Kikuchi, M., Watase, S., Izaki, M., Naito, H., Hiramoto, M.: Hall effect in bulk-doped organic single-crystal. Adv. Mater. **29**, 1605619 (6 pages) (2017)
13. Kallman, H., Pope, M.: In: Kallman H., Silver, M. (eds.) Proceedings of Symposium on Electrical Conductivity in Organic Solids, pp. 1–2. Interscience Wiley, New York. Duke University, Durham, North Carolina, 20–22 April (1960)
14. Kloc, Ch., Simpkins, P.G., Siegrist, T., Laudise, R.A.: Physical vapor growth of centimeter-sized crystals of α-hexathiophene. J. Crystal Growth **182**, 416 (1997)
15. Laudise, R.A., Kloc, Ch., Simpkins, P.G., Siegrist, T.: Physical vapor growth of organic semiconductors. J. Crystal Growth **187**, 449 (1998)
16. Wagner, H.J., Loutfy, R., Hsiao, C.-K.: Purification and characterization of phthalocyanines. J. Mater. Sci. **17**, 2781–2791 (1982)
17. Hiramoto, M., Miki, A., Yoshida, M., Yokoyama, M.: Photocurrent multiplication in organic single-crystals. Appl. Phys. Lett. **81**, 1500–1502 (2002)
18. Frenkel, J.: On pre-breakdown phenomena in insulators and electronic semiconductors. Phys. Rev. **54**, 647–648 (1938)
19. Simmons, J.G.: Poole-Frenkel effect and Schottky effect in metal-insulator-metal systems. Phys. Rev. **155**, 657–660 (1967)
20. No tiitle
21. Harrell, W.R., Frey, J.: Observation of Poole-Frenkel effect saturation in SiO_2 and other insulating films. Thin Solid Films **352**, 195–204 (1999)
22. Adachi, C.: Device physics of organic semiconductors (Chap. 2), p. 17. Kodansha Scientific (2012) (in Japanese)
23. Fowler, R.H., Nordheim, L.: Electron emission in intense electric fields. Proc. R. Soc. Lond. Ser. A **119**, 173–181 (1928)
24. Simmons, J.O.: Generalized formula for the electric tunnel effect between similar electrodes separated by a thin insulating film. Appl. Phys. Lett. **34**, 1793–1803 (1963)
25. Parker, I.D.: Carrier tunneling and device characteristics in polymer light-emitting diodes. J. Appl. Phys. **75**, 1656–1666 (1994)
26. Sze, S.M.: Physics of Semiconductor Devices, p. 552. Wiley, New York (1981)
27. Tong, Y., Fuhr, J.D., Martiarena, M.L., Oughaddou, H., Enriquez, H., Nicolas, F., Chaouchi, K., Kubsky, S., Bendounan, A.: Properties of NTCDA thin films on Ag(110): scanning tunneling microscopy, photoemission, near-edge X-ray fine structure, and density functional theory investigations. J. Phys. Chem. C **123**, 379–386 (2019)

28. Hiramoto, M., Kikuchi, M., Izawa, S.: Parts-per-million-level doping effects in organic semiconductor films and organic single-crystals. Adv. Mater. **30**, 1801236 (15 pages) (2018)
29. Zeng, X., Wang, L., Duan, L., Qiu, Y.: Homoepitaxy growth of well-ordered rubrene thin films. Cryst. Growth Design. **8**, 1617–1622 (2008)
30. Schift, H.: Nanoimprint lithography: 2D or not 2D?A review. Appl. Phys. A **121**, 415–435 (2015)
31. Chou, S.Y., Krauss, P.R., Renstrom, P.J.: Imprint of sub25 nm vias and trenches in polymers. Appl. Phys. Lett. **67**, 3114–3116 (1995)
32. Hirai, Y.: Polymer science in nanoimprint lithography. J. Photopolym. Sci. Tech. **18**, 551–558 (2005)
33. Sreenivasan, S.V.: Nanoimprint lithography steppers for volume fabrication of leading-edge semiconductor integrated circuits. Microsyst. Nanoeng. **3**, 17075 (19 pages) (2017)

Chapter 6
Photocurrent Multiplication at Organic Heterojunctions

6.1 Background

In Chapters from 2 to 5, the photocurrent multiplication occurring at the organic/metal interface [1–3] has been summarized. The multiplication rate reaches 10^5-fold at room temperature [3]. The multiplication is caused by the tunneling electron injection from a metal electrode to organic semiconductors, triggered by trapped holes near the interface [1–3]. As carrier traps, a molecular blind alley model (structural trap model) has been proposed [4–7].

6.2 Motivation

The structural traps formed by the spatial gaps and molecular blind alleys are the primary cause of photocurrent multiplication at the organic/metal interface. Similarly, such structural traps should also form at the organic heterointerface, which is the motivation for this investigation on multiplication phenomena at organic heterojunctions. This idea is supported by the reasoning that spatial gaps might exist at the heterojunction interface between different organic polycrystalline films, because organic microcrystals of different types of organic semiconductors have the tendency to aggregate separately from each other. Additionally, molecular steps and kinks (molecular blind alleys) inevitably exist on the surface of organic microcrystals.

M. Hiramoto, *Organic Photocurrent Multiplication*, Electronic Materials: Science & Technology, https://doi.org/10.1007/978-981-99-1237-7_6

6.3 Photocurrent Multiplication at CuPc/Me-PTC Heterojunction

A photocurrent multiplication, with the rate reaching 3,000-fold, was first achieved at an organic heterojunction [7, 8]. The photocurrent multiplication caused by red light irradiation was suppressed by superimposed blue light irradiation.

6.3.1 CuPc/Me-PTC Heterojunction

As organic heterojunctions, we selected *p*-type copper phthalocyanine (CuPc) [9] and *n*-type perylene pigment (Me-PTC) [10–12]. CuPc (Fig. 6.1) and Me-PTC (Fig. 6.1), purified twice by train sublimation [13], were used. The structure of the heterojunction cell is shown in Fig. 6.1. CuPc film of thickness 600 nm and Me-PTC film of thickness 100 nm were successively deposited at a rate of 0.1 nms^{-1} on an indium-tin oxide (ITO) glass substrate, by vacuum evaporation under a pressure of 1×10^{-3} Pa. A semitransparent Au electrode of 20 nm thickness was deposited on the Me-PTC surface.

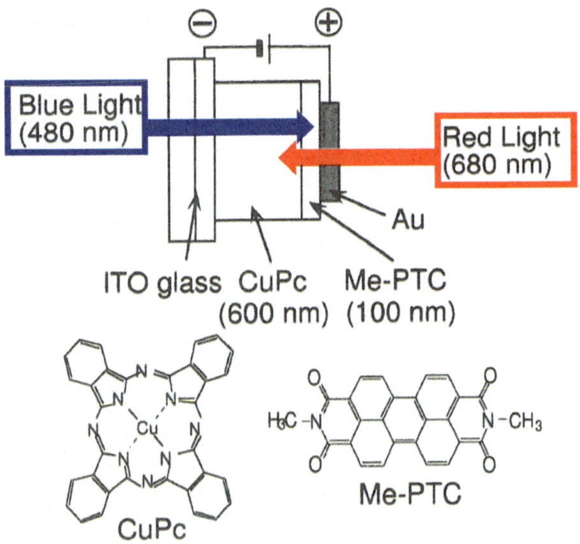

Fig. 6.1 Structure of the organic heterojunction cell. Thicknesses of CuPc, Me-PTC, and Au films are 600, 100, and 20 nm, respectively. ITO electrode was negatively biased with respect to Au electrode. Chemical structures of CuPc and Me-PTC are also shown. Reproduced with the permission of [7]. © 1997 Elsevier Science S. A.

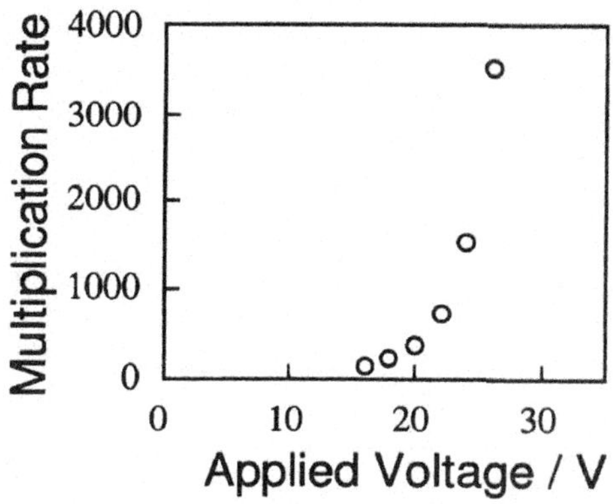

Fig. 6.2 Dependence of multiplication rate on applied voltage. Au electrode was positively biased with respect to ITO electrode. Monochromatic light of wavelength 680 nm used to selectively excite CuPc was irradiated on the Au electrode. Measurements were carried out at a pressure of 10^{-1} Pa, at room temperature. Reproduced with the permission of [7]. © 1997 Elsevier Science S. A.

6.3.2 Multiplication Rate

The dependence of multiplication rate on applied voltage, as observed for a CuPc/Me-PTC heterojunction cell (Fig. 6.1), is shown in Fig. 6.2. The multiplication rate, which is, the photocurrent quantum efficiency, was calculated as a ratio of the number of carriers flowing through the device by red-light illumination to the number of photons absorbed by the CuPc film. When the Au electrode was biased positively, carrier injection from both metal electrodes in the dark was effectively prevented, and the photocurrent multiplication phenomenon was dominant with the rate exceeding 3,000-fold.

6.3.3 Identification of Interface

6.3.3.1 Action Spectrum

The active interface can be identified from the photocurrent action spectrum in the region where photons are absorbed by the CuPc film. The ITO electrode was biased negatively by irradiation using a monochromatic light. The multiplied photocurrent showed maxima in the absorption edges of the CuPc film (around 560, 760 nm) and a bottom between two absorption peaks (660 nm), where the incident photons can penetrate deeply into the CuPc film (Fig. 6.3). On the contrary, the multiplied photocurrent showed minima in the wavelength region of absorption peaks (around 620 and 700 nm), where most of the incident photons were absorbed near the interface of the negatively biased ITO electrode. However, when the positively biased Au electrode was illuminated, the peaks were no longer distinctly visible. These results suggest that the CuPc/Me-PTC heterointerface or the Me-PTC/Au interface is responsible for the observed multiplication process.

6.3.3.2 Deposition Rate

When the deposition rate of organic layers was kept at a constant value of 0.1 nms^{-1}, as in the case of ITO/CuPc (thickness: 600 nm, deposition rate: 0.1 nms^{-1})/Me-PTC (100 nm, 0.1 nms^{-1})/Au cell, large multiplication was observed (Fig. 6.2). However, when only a very thin CuPc layer (50 nm) neighboring the CuPc/Me-PTC heterojunction was evaporated at a 40 times higher deposition rate of 4 nms^{-1}, as in the case of the ITO/CuPc (550 nm, 0.1 nms^{-1})/CuPc (50 nm, 4 nms^{-1})/Me-PTC (100 nm, 0.1 nms^{-1})/Au cell, small multiplication was observed. Based on this result, the interface responsible for the photocurrent multiplication was identified as the CuPc/Me-PTC heterojunction.

6.3.4 Multiplication Suppression by Superimposed Light

The CuPc and Me-PTC layers were excited simultaneously by two lights of different wavelength (Fig. 6.1). CuPc is selectively excited by irradiation using red light of wavelength 680 nm (Fig. 6.3), and Me-PTC is selectively excited by irradiation using blue light of wavelength 480 nm, from the opposite direction (see Fig. 6.1), due to the absence of CuPc absorption (450–510 nm) (Fig. 6.3). This effectively suppresses the photocurrent multiplication (Fig. 6.4). This response was reversible, and observed in the wavelength region of the superimposed second light (400–540 nm), which corresponds to the absorption of the Me-PTC film (Fig. 6.5).

6.3.5 Multiplication Mechanism

The interface causing photocurrent multiplication was identified to be the CuPc/Me-PTC heterojunction. Rise of multiplied photocurrent after red-light on and after blue-light off exhibited significant slow transient response on the order of several tens of seconds (Fig. 6.4). This indicates a close relationship between the trapping of photogenerated carriers and photocurrent multiplication. Based on this observation, we propose the following multiplication mechanism, assuming neutral electron trap near the CuPc/Me-PTC heterojunction.

The energy structure of the cell during photocurrent multiplication is shown in Fig. 6.6a. When CuPc is excited by red light, the electrons and holes photogenerated in the CuPc layer, move along the potential gradient. Some of the photogenerated electrons reaching the CuPc/Me-PTC heterojunction are captured by the traps at the heterojunction. Accumulation of trapped electrons at the interface results in a high

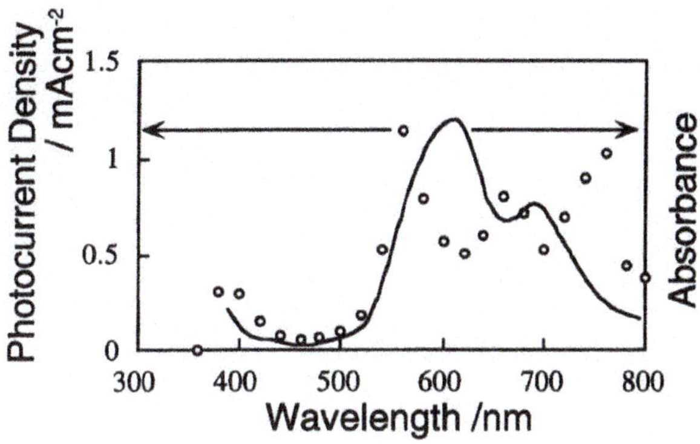

Fig. 6.3 Spectral dependence of multiplied photocurrent (open dots) for a cell in Fig. 6.1. ITO electrode was negatively biased (20 V) with respect to the Au electrode by irradiation using a monochromatic light. Absorption spectrum of CuPc film (solid curve) is also shown. Reproduced with the permission of [7]. © 1997 Elsevier Science S. A.

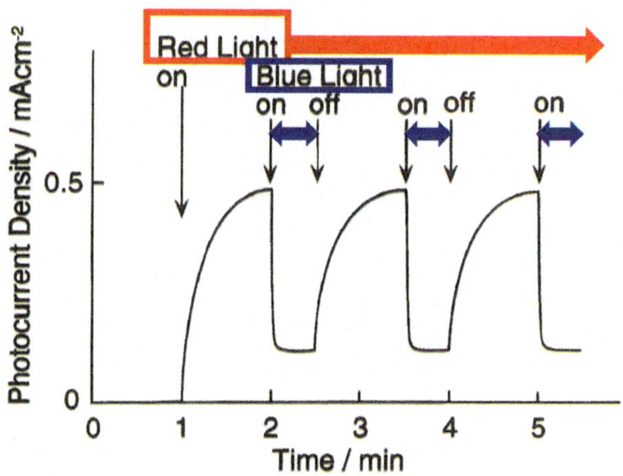

Fig. 6.4 Typical suppression of photocurrent multiplication by blue light (480 nm) irradiation on the ITO electrode. Red light (680 nm) was always irradiated on the Au electrode. Reproduced with the permission of [8]. © 1998 Elsevier Science S. A.

electric field across the Me-PTC layer. Due to this high energy field, the energy band is strongly inclined causing the tunneling injection of electrons from the valence band of CuPc to the conduction band of Me-PTC (Fig. 6.6a, yellow arrow), resulting in a 3000-fold photocurrent multiplication rate.

6.3.6 Suppression Mechanism

The proposed mechanism for multiplication suppression is shown in Fig. 6.6b. Extinction of accumulated electrons occurs due to the recombination with holes

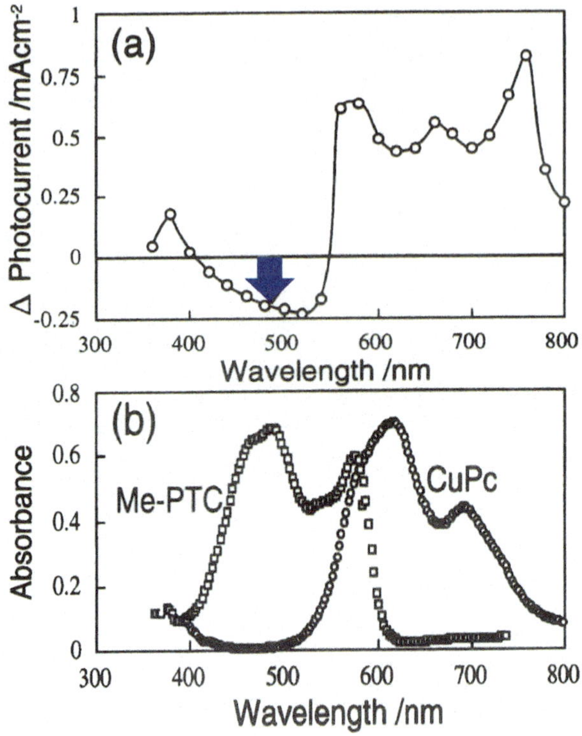

Fig. 6.5 **a** Spectral dependence of the change in photocurrent multiplication under the simultaneous irradiation of red light (680 nm) on the Au electrode and blue light on the ITO electrode. Solid line at zero shows the level of photocurrent multiplication when the red light alone is turned on. The multiplication rate at this level is approximately 300-fold. **b** Absorption spectra of Me-PTC (100 nm) and CuPc (100 nm) films. Reproduced with the permission of [8]. © 1998 Elsevier Science S. A.

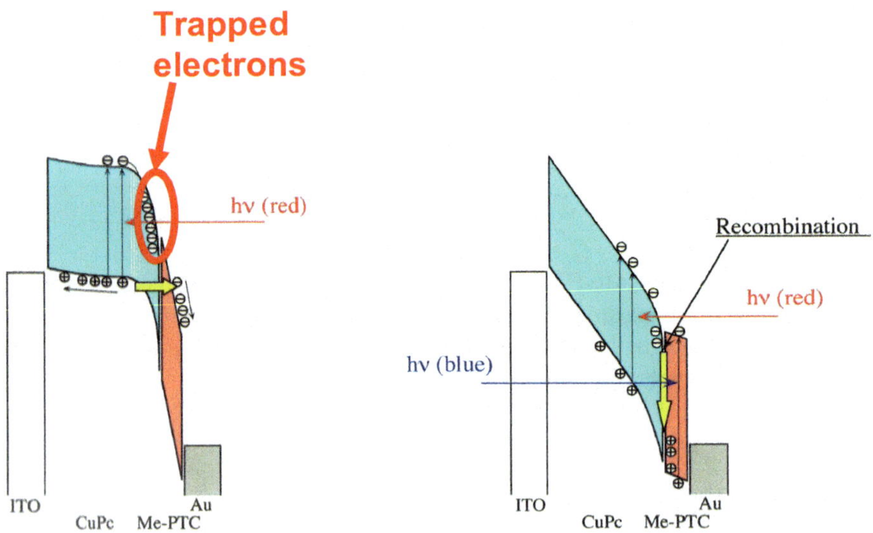

Fig. 6.6 Energy structures of ITO/CuPc/Me-PTC/Au cell under an applied electric field: **a** photocurrent multiplication under red light irradiation. **b** Multiplication suppression by blue light superimposed on red light. Reproduced with the permission of [8]. © 1998 Elsevier Science S. A.

photogenerated by the superimposed blue light in the Me-PTC film. This effectively suppresses the multiplication.

The photocurrent suppression by the blue light exciting Me-PTC strongly supports the mechanism that electrons are trapped at the location of CuPc/Me-PTC heterojunction. If the electron traps were located at Me-PTC/Au interface, multiplication suppression by Me-PTC excitation cannot be explained.

6.3.7 Structural Trap at Organic Heterojunction

In the Chaps. 3, 4 and 5, we discussed structural traps at organic/metal interface, which give rise to the photocurrent multiplication (Chap. 5, Fig. 5.13). The spatial gap and the molecular blind alleys(molecular steps on the crystalline surface), compose the structural traps. At a high deposition rate of CuPc, its crystallization is inhibited, resulting in the formation of amorphous CuPc layer (50 nm) adjacent to Me-PTC layer. As a result, molecular blind alleys on the CuPc surface located at CuPc/Me-PTC heterojunction are not formed and the multiplication process is inhibited (see Sect. 6.3.3.2).

Based on these results, the following can be inferred about the structural traps:

(i) Traps are located at the CuPc side of the CuPc/Me-PTC heterojunction.
(ii) Traps exist only on the surface of organic microcrystals.

Therefore, we suppose that structural traps similar to the organic/metal junctions are formed at the organic heterojunctions. The schematic illustration of structural traps at CuPc/Me-PTC heterojunction is shown in Fig. 6.7. Crystalline CuPc and Me-PTC films consist of a vast number of microcrystals that are several hundred nanometers in size. The molecular steps and kinks on the surface of the microcrystals act as molecular blind alleys. In addition, there are spatial gaps between the CuPc and Me-PTC microcrystals, as different kinds of organic microcrystals aggregate separately from each other. These structural traps give rise to the large multiplication at the organic heterojunction.

6.4 Organic Heterojunctions Incorporating Hole and Electron Transporting Layers

Both organic heterojunctions, the one between the hole transporting layer (HTL) and *n*-type organic semiconductor (*n*-OSC) layer (Fig. 6.8a), as well as the one between the electron transporting layer (ETL) and *p*-type organic semiconductor (*p*-OSC) layer (Fig. 6.8b), showed photocurrent multiplication. The multiplication mechanism exhibited a front and back relationship, that is the multiplication is caused by the trapped holes in the former and by the trapped electrons in the latter.

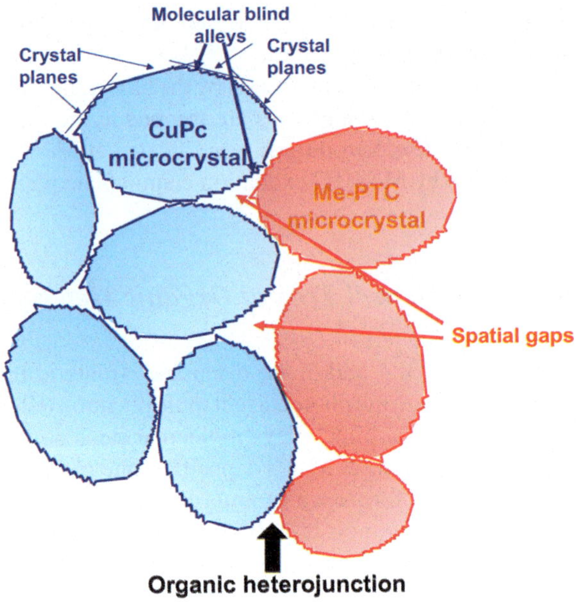

Fig. 6.7 Schematic illustration of CuPc/Me-PTC interface. Both CuPc and Me-PTC form polycrystalline films consisting of a vast number of microcrystals that are several hundred nanometers in size. The molecular steps and kinks on the surface of the crystals act as molecular blind alleys. In addition, there are spatial gaps between CuPc and Me-PTC microcrystals

6.4.1 Multiplication Rate

The cell structure and voltage dependence of the multiplication rate for HTL/*n*-OSC heterojunction are shown in Fig. 6.8a. A transparent TPD (Fig. 6.8a) layer (50 and 100 nm) [14, 15], and an Me-PTC layer (600 nm) were used as HTL and *n*-OSC, respectively. The TPD layer was biased negatively with respect to the Me-PTC layer. By exciting the Me-PTC using a monochromatic light of wavelength 480 nm, the multiplication rate reached 2,000-fold at 20 V. Increasing the TPD thickness from 50 to 100 nm suppressed the multiplication indicating that the electric field concentration occurs mainly across the HTL (which acts as an insulating layer).

The cell structure and voltage dependence of multiplication rate for ETL/*p*-OSC heterojunction, are shown in Fig. 6.8b. A transparent OXD layer (50 nm) [16] and quinacridone pigment (DQ) (600 nm) [2, 17, 18] were used as ETL and *p*-OSC, respectively. The OXD layer was biased positively with respect to the DQ layer. By exciting DQ using a monochromatic light of wavelength 580 nm, the multiplication rate exceeded 300-fold at 50 V.

The bias polarity causing the photocurrent multiplication at the TPD/Me-PTC heterojunction and the OXD/DQ heterojunction, were opposite.

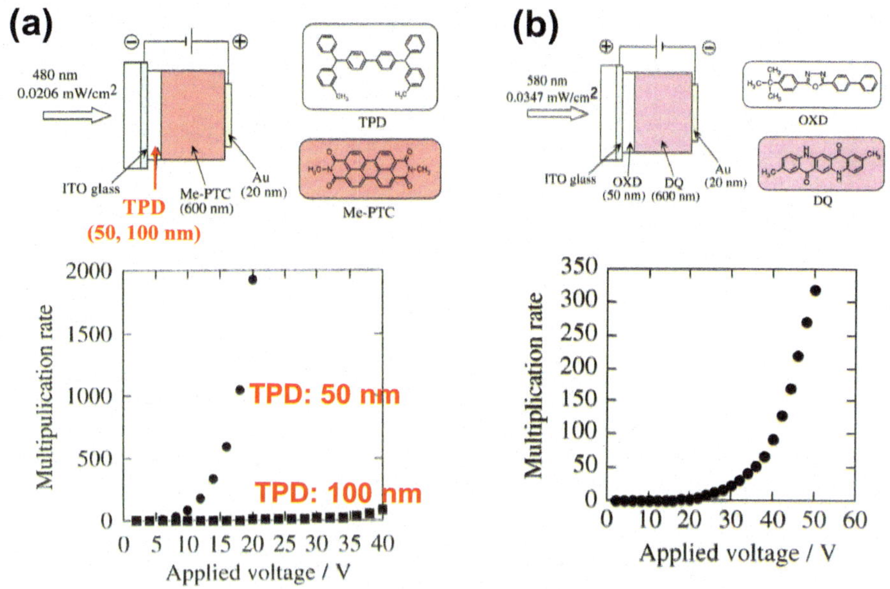

Fig. 6.8 **a** Cell structure and voltage dependence of multiplication rate for TPD (50, 100 nm)/*n*-Me-PTC (600 nm) heterojunction. The transparent TPD layer was biased negatively with respect to Me-PTC layer. Me-PTC was excited via irradiation of the monochromatic light of wavelength 480 nm on the TPD side. **b** Cell structure and the voltage dependence of multiplication rate for OXD (50 nm)/*p*-DQ (600 nm) heterojunction. The OXD layer was biased positively with respect to DQ layer. DQ was excited via irradiation of the monochromatic light of wavelength 580 nm on OXD

6.4.2 Action Spectra

The action spectra of the multiplied photocurrent for the TPD/Me-PTC heterojunction, is shown in Fig. 6.9a. Upon irradiation of the transparent TPD side with monochromatic light (Fig. 6.9a, left), the shape of the action spectrum (black dots) coincided well with the absorption spectrum of the Me-PTC (solid curve) (symbatic relationship). In contrast, upon irradiation of the Me-PTC side with monochromatic light (Fig. 6.9a, right), the shape of the action spectrum (black dots) showed an antibatic relationship with the absorption spectrum of Me-PTC (solid curve), that is sharp peaks appear at the edges of the absorption spectrum and a dip appears in the absorption spectrum where the absorbance is low. Based on this, it can be inferred that the light penetrating the bulk of the Me-PTC layer and reaching the TPD/Me-PTC heterointerface is what gives rise to the large photocurrent multiplication. Thus, the multiplication process was confirmed to occur at the TPD/Me-PTC heterointerface.

The action spectra of the multiplied photocurrent for the OXD/DQ heterojunction, is shown in Fig. 6.9b. Upon irradiation of the transparent OXD side (Fig. 6.9b, left) and DQ side (Fig. 6.9b, right), it was observed that the relationship between action spectrum (black dots) and absorption spectrum of DQ (solid curve) were symbatic

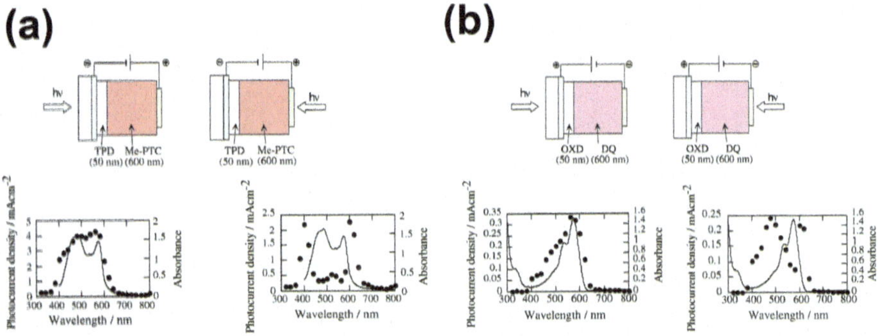

Fig. 6.9 **a** Action spectra of the multiplied photocurrent density for TPD/Me-PTC heterojunction. Monochromatic light was irradiated on the TPD side (left) and on Me-PTC side (right). Absorption spectrum of Me-PTC is also shown (solid curves). TPD is transparent in the visible region. **b** Monochromatic light was irradiated on OXD side (left) and on DQ side (right). Absorption spectrum of DQ is also shown (solid curves). OXD is transparent in the visible region

and antibatic, respectively. Thus, the multiplication process was confirmed to occur at the OXD/DQ heterointerface.

6.4.3 *Multiplication Mechanism*

The characteristics of the photocurrent multiplication observed for organic heterojunctions are summarized as follows.

(i) Photocurrent multiplication occurs at organic heterojunctions, irrespective of whether it is the HTL/*n*-OSC or ETL/*p*-OSC heterojunction.
(ii) The bias polarity causing photocurrent multiplication was opposite for HTL/*n*-OSC and ETL/*p*-OSC heterojunctions, that is, (−)TPD/Me-PTC(+) and (+)OXD/DQ(−).
(iii) The electric field concentration occurs mainly across the HTL (or ETL), acting as an insulating layer.

Based on the analogy of photocurrent multiplication at organic/metal interface (Chaps. 2, 3, 4 and 5), we propose the following multiplication mechanism.

The energy structure during multiplication when the TPD layer is biased negatively with respect to the Me-PTC layer is shown in Fig. 6.10a. Light irradiation of Me-PTC, causes the photogenerated holes to move toward the TPD/Me-PTC heterojunction, and are captured by traps on Me-PTC surface at TPD/Me-PTC interface. This accumulation of trapped holes results in the electric field concentration across the insulating TPD. Finally, the tunneling injection of electrons from the valence band of TPD to the conduction band of Me-PTC occurs.

The energy structure during multiplication when the OXD layer is biased positively with respect to DQ layer, is shown in Fig. 6.10b. Light irradiation of DQ,

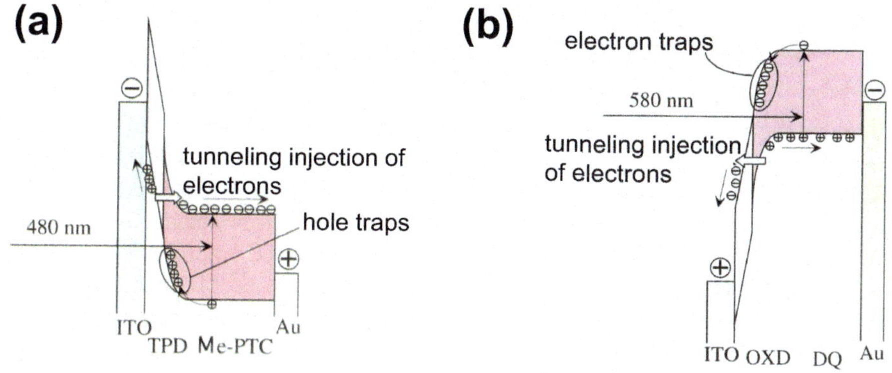

Fig. 6.10 **a** Energy structure during multiplication for TPD/Me-PTC heterojunction. **b** Energy structure during multiplication for OXD/DQ heterojunction

causes the photogenerated electrons to move toward the OXD/DQ heterojunction, and are captured by traps on DQ surface at the OXD/DQ interface. This accumulation of trapped electrons results in the electric field concentration across the insulating OXD. Finally, the tunneling injection of electrons from the valence band of DQ to the conduction band of OXD occurs.

A Comparison of the multiplication between HTL/*n*-OSC and ETL/*p*-OSC is summarized in Table 6.1. The bias polarity is opposite for HTL/*n*-OSC and ETL/*p*-OSC, and the carrier trap causing multiplication are holes and electrons for HTL/*n*-OSC and ETL/*p*-OSC, respectively. Both HTL and ETL, where the electric field is concentrated, are insulating in nature. Directions of the tunneling injection of electrons are also opposite, that is from HTL to *n*-OSC and from *p*-OSC to ETL, for HTL/*n*-OSC and ETL/*p*-OSC heterojunctions, respectively.

Table 6.1 Characteristics of multiplication for HTL/*n*-OSC and ETL/*p*-OSC heterojunctions

	HTL/*n*-OSC	ETL/*p*-OSC
Bias polarity	(−) HTL / *n*-OSC (+)	(+) ETL / p-OSC (−)
Carrier trap	Hole trap	Electron trap
Field concentration layer	HTL	ETL
Direction of tunneling injection of electrons	HTL → *n*-OSC	*p*-OSC → ETL

6.4.4 Structural Traps at HTL/n-OSC and ETL/p-OSC HTL (ETL)/n-OSC (p-OSC) Heterojunctions

A schematic illustration of the TPD/Me-PTC interface is shown in Fig. 6.11. TPD forms molecularly smooth amorphous film [14, 15, 19] and *n*-Me-PTC forms a polycrystalline film consisting of Me-PTC microcrystals [6]. Molecular steps and kinks on the surface of the Me-PTC microcrystals act as molecular blind alleys. In addition, spatial gaps are formed between the Me-PTC microcrystals and the amorphous TPD. As a result, molecular blind alleys on the Me-PTC surface act as structural traps (Chap. 5) and give rise to the large photocurrent multiplication. Similar structural traps were also formed in the OXD/DQ heterojunction.

It should be noted that there are energy barriers for holes from the valence band of Me-PTC (5.4 eV) [6] to that of TPD (5.6 eV) [20], and for electrons from the conduction bands of DQ (3.4 eV) [2, 17] to that of OXD (2.5 eV). Thus, there is a possibility that the carrier accumulation is a result of these energy barriers.

While the nature of structural traps at organic/metal junctions has been described in Chap. 5, a detailed study will be required to understand the nature of structural traps formed at various types of organic heterojunctions.

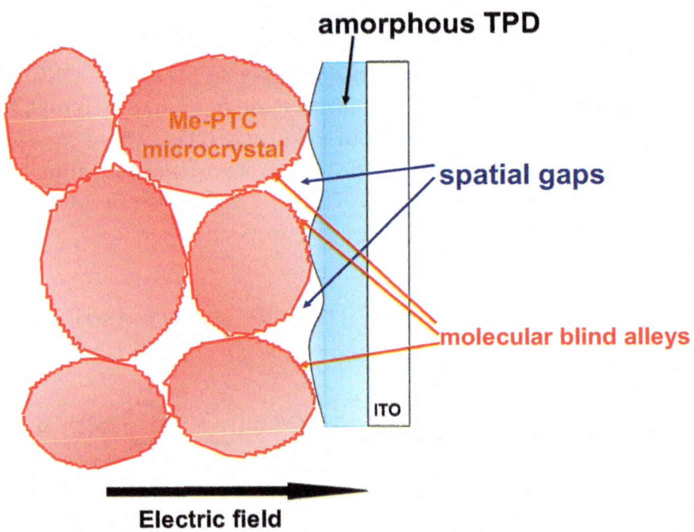

Fig. 6.11 Schematic illustration of TPD/Me-PTC interface. Me-PTC forms polycrystalline film gathering of a lot of microcrystals that are several hundred nanometers in size. TPD forms amorphous film. There are the spatial gaps between amorphous TPD and polycrystalline Me-PTC. There are a lot of molecular steps and kinks on the Me-PTC microcrystal surfaces, which act as molecular blind alleys

6.5 Conclusion

Photocurrent multiplication was first achieved in organic heterojunctions such as *p*-CuPc/*n*-Me-PTC, HTL/*n*-OSC, and ETL/*p*-OSC. The photocurrent multiplication caused by the irradiating red light was effectively suppressed by the superimposed irradiating blue light for *p*-CuPc/*n*-Me-PTC heterojunction. The multiplication mechanism of HTL/*n*-OSC and that of ETL/*p*-OSC show front and back relationship. A structural trap model at organic heterojunctions was proposed.

References

1. Hiramoto, M., Imahigashi, T., Yokoyama, M.: Photocurrent multiplication in organic pigment films. Appl. Phys. Lett. **64**, 187–189 (1994)
2. Hiramoto, M., Kawase, S., Yokoyama, M.: Photoinduced hole injection multiplication in p-type quinacridone pigment films. Jpn. J. Appl. Phys. **35**, L349–L351 (1996)
3. Katsume, T., Hiramoto, M., Yokoyama, M.: Photocurrent multiplication in naphthalene tetracarboxylic anhydride film at room temperature. Appl. Phys. Lett. **69**, 3722–3724 (1996)
4. Hiramoto, M., Nakayama, K., Katsume, T., Yokoyama, M.: Field-activated structural traps at organic pigment/metal interfaces causing photocurrent multiplication phenomena. Appl. Phys. Lett. **73**, 2627–2629 (1998)
5. Hiramoto, M.: Molecular-sized structural trap at organic-metal interface and photocurrent multiplication phenomenon. In: Salaneck, W.R., Seki, K., Kahn, A., Pireaux, J.-J. (eds.) Conjugated Polymer and Molecular Interfaces (Chap. 18), pp. 585–612. Marcel Dekker Inc., New York/Basel (2002), and references therein
6. Nakayama, K., Hiramoto, M., Yokoyama, M.: Photocurrent multiplication at organic/metal interface and surface morphology of organic films. J. Appl. Phys. **87**, 3365–3369 (2000)
7. Hiramoto, M., Kumaoka, H., Yokoyama, M.: Photocurrent multiplication phenomenon at organic/organic heterojunction and application to optical computing device combining with organic electroluminescence. Synth. Met. **91**, 77–79 (1997)
8. Hiramoto, M., Nakayama, K., Sato, I., Kumaoka, H., Yokoyama, M.: Photocurrent multiplication phenomena at organic/metal and organic/organic interfaces. Thin Solid Films **331**, 71–75 (1998)
9. Tang, C.W.: Two-layer organic photovoltaic cell. Appl. Phys. Lett. **58**, 1062–1064 (1986)
10. Hiramoto, M., Kishigami, Y., Yokoyama, M.: Doping effect on the two-layer organic solar cell. Chem. Lett. **1990**, 119–122 (1990)
11. Hiramoto, M., Suezaki, M., Yokoyama, M.: Effect of thin gold interstitial-layer on the photovoltaic properties of tandem organic solar cell. Chem. Lett. **1990**, 327–330 (1990)
12. Hiramoto, M., Fujiwara, H., Yokoyama, M.: Three-layered organic solar cell with a photoactive interlayer of codeposited pigments. Appl. Phys. Lett. **58**, 1062–1064 (1991)
13. Wagner, H.J., Loutfy, R.O., Hsiao, C.: Purification and characterization of phthalocyanines. J. Mater. Sci. **17**, 2781–2791 (1982)
14. Vamvounis, G., Aziz, H., Hu, N., Popovic, Z.D.: Temperature dependence of operational stability of organic light emitting diodes based on mixed emitter layers. Synth. Met. **143**, 69–73 (2004)
15. Shirota, Y., Kuwabata, Y., Inada, H., Wakimoto, T., Nakada, H., Yonemoto, Y., Kawami, S., Imai, K.: Multilayered organic electroluminescent device using a novel starburst molecule 4,4',4''-tris(3-methylphenylamino)triphenylamine, as a hole transport material. Appl. Phys. Lett. **65**, 807–809 (1994)

16. Hamada, Y., Adachi, C., Tsutsui, T., Saito, S.: Blue-light-emitting organic electroluminescent devices with oxadiazole dimer dyes as an emitter. Jpn. J. Appl. Phys. **31**, 1812–1816 (1992)
17. Hiramoto, M., Nakamura, H., Shiojiri, T., Yokoyama, M.: Doping effects of organic solar cells using quinacridone pigment. Kobunshi Ronbunshu **47**, 915–920 (1990). [in Japanese]
18. Wang, C., Zhang, Z., Wang, Y.: Quinacridone-based π-conjugated electronic materials. J. Mater. Chem. C. **4**, 9918–9936 (2016) and references therein
19. Han, E., Do, L., Niidome, Y., Fujihira, M.: Observation of crystallization of vapor-deposited TPD films by AFM and FFM. Chem. Lett. **1994**, 969–972 (1994)
20. Ito, E., Oji, H., Hayashi, N., Ishii, H., Ouchi, Y., Seki, K.: Electronic structures of TPD/metal interfaces studied by photoemission and Kelvin probe method. Appl. Surf. Sci. **175–176**, 407–411 (2001)

Chapter 7
High-Speed Response Devices

7.1 Background

7.1.1 Transient Response of Photocurrent Multiplication

The typical transient response of photocurrent multiplication described in Chap. 3 is shown in Fig. 7.1a [1, 2]. The transient photocurrent consisted of two components. The first component (red circle) rose within 10 μs and reached a plateau. This is the primary photocurrent due to carrier photogeneration in the organic semiconductor film (Fig. 7.1b). The photocurrent quantum efficiency of the first component in the plateau region was 0.65% for the Me-PTC film (Chap. 3). The second component (blue circle) rising after several tens of milliseconds of light irradiation is the multiplied photocurrent owing to the tunneling injection of electrons, which is caused by hole accumulation (Fig. 7.1c).

7.1.2 Structural Trap Model

The nanostructure of the NTCDA/Au interface, illustrated on a real scale, is shown in Fig. 7.2. First, there were several spatial gaps behind the 20 nm spherical Au nanoparticles. Second, molecular steps with a spatial depth of 0.5 nm act as molecular blind alleys. A detailed discussion is presented in Sect. 7.5.

Photogenerated holes reach the interface and are captured by molecular steps that serve as blind alleys. To escape from the molecular blind alley, the holes should experience a concentrated electric field (curved red arrows). As a result, the delivery of holes to the Au electrode is suppressed, and the holes accumulate in molecular blind alleys under the spatial gaps. Finally, more electrons are injected by tunneling from the contact sites of the Au nanoparticles (blue arrows). This mechanism leads to photocurrent multiplication.

M. Hiramoto, *Organic Photocurrent Multiplication*, Electronic Materials: Science & Technology, https://doi.org/10.1007/978-981-99-1237-7_7

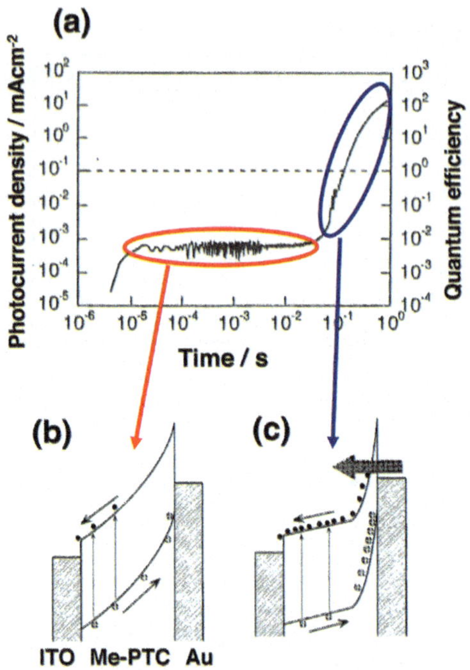

Fig. 7.1 **a** Transient response of the photocurrent observed in ITO/Me-PTC/Au device. Bias polarity is (+)/ITO/Me-PTC/Au(−). Red and blue circles corresponds to the first component, i.e., the primary photocurrent and the second component, i.e., the multiplied photocurrent injected by the tunneling. **b** Energy structure under primary photocurrent flow (red circle). **c** Energy structure under multiplied photocurrent flow (blue circle). Reproduced with the permission of [1]. © 1998 American Institute of Physics

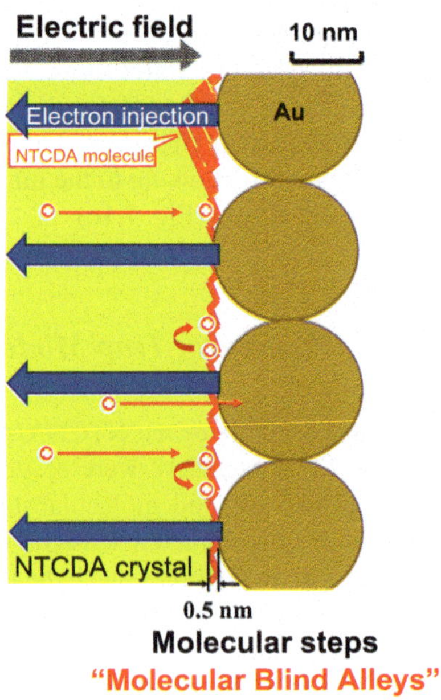

Fig. 7.2 The nanostructure of the NTCDA/Au interface illustrated on a real scale. The molecular blind alleys act as structural traps

7.2 Numerical Calculation

The key for constructing high-speed response devices are the increase in quantum efficiency of primary photocurrent and the quick establishment of a steady state for the occupation of molecular blind alleys.

7.2.1 Motivation

The slow photoresponse reaching several tens of seconds of multiplied photocurrent was the main hurdle in constructing practical organic multiplication-type photodetectors. To clarify the keys to construct high-speed response devices, numerical calculation based on the structural trap model (Fig. 7.2) [3, 4] was performed.

7.2.2 Simplified Model for Calculation

For the numerical calculation, we used a simplified structural trap model (Fig. 7.3) based on the structural trap model (Fig. 7.2) [5, 6]. There were spatial gaps between the metal nanoparticles with diameter 20 nm. In this model, holes are trapped by structural traps and electrons are injected from the metal electrode. The photogenerated holes were continuously supplied from the bulk of organic film. The photogenerated carriers were assumed to disappear only at the contact points between the organic surface and the metal nanoparticles. We have introduced bulk mobility and surface mobility previously. Because there are a lot of molecular blind alleys that capture holes under concentrated electric field, we expected the surface mobility would be lower than the bulk mobility.

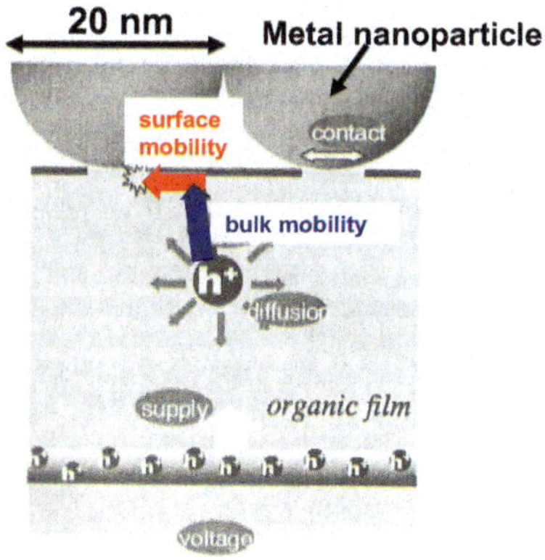

Fig. 7.3 Simplified structural trap model for numerical calculation

Numerical calculations were performed to simulate the time-developing charge distribution at the organic/metal junction and the dynamic flow of the photogenerated holes (Fig. 7.3). At a certain time step, each hole within the organic semiconductor film was moved along the calculated electric field. For the hole movement, the field drift and diffusion drift were considered. The time development of the hole distribution was obtained sequentially. Photogenerated holes were supplied to the film bulk at each time step.

7.2.3 *Time Development of Hole Accumulation*

Figure 7.4 shows the typical time development of hole distribution in the simplified model (Fig. 7.3) under a continuous supply of photogenerated holes from the bulk of organic film. The photogenerated holes reaching the surface of the organic film apparently pile up at the non-contact region between the two metal nanoparticles at a high density of surface charges of holes until they vanish at the contact point. In a steady-state, such hole accumulation at the surface is balanced with the supply of photogenerated holes from the bulk and vanishing at the metal electrode. Thus, the excess space charge remaining at the film surface causes electron injection from the metal electrode.

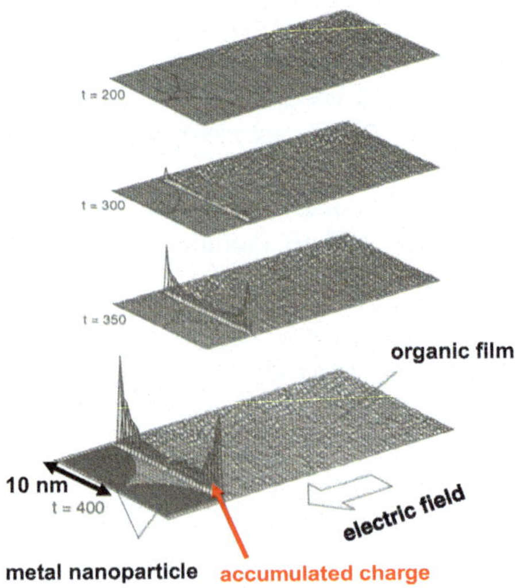

Fig. 7.4 Time development of the charge distribution at the organic/metal interface under continuous light irradiation

7.2.4 Surface Mobility

When the bulk mobility and surface mobility (Fig. 7.3) were the same, however, the amount of hole accumulation at the organic film surface was limited by the short staying time of the holes at the organic surface, and the steady state was established quickly, that is, the response speed was very fast, although the multiplication rate was very small. This contradicts the fact that the multiplied photocurrent usually exhibits a very slow response (Fig. 7.5a).

Based on the effect of molecular blind alleys on the surface of organic microcrystals (Fig. 7.2), the hole movement along the crystal surface should be obstructed. The surface mobility (Fig. 7.3, red arrow) should be lower than that of the bulk (Fig. 7.3, blue arrow). Low surface mobility is an expression of the trapping effect of molecular blind alleys on the organic crystal surface. Thus, we introduced low surface mobility to explain the slow response.

Figure 7.5b shows the transient response of the tunneling injection current (second component) calculated from the time of developing the electric field at the contact point with various mobility ratios. The surface mobility ratio is defined as the ratio of the surface mobility to the bulk mobility. Interestingly, introducing the concept of low surface mobility reproduced the multiplied current transients that reached 1 s after the onset time (second component).

The response speed corresponds to the time required to reach a steady state of the number of holes captured by the molecular blind alleys. At steady state, the hole accumulation rate to the molecular blind alleys at the non-contact area of the organic surface and the hole release rate to the metal electrode were equilibrated. Numerical

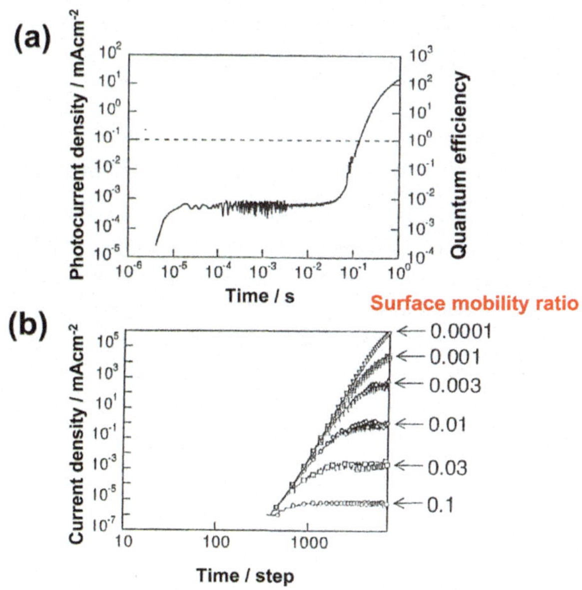

Fig. 7.5 **a** Transient response of the photocurrent observed in (+)ITO/Me-PTC/Au(−) device experimentally. **b** Simulated responses of the secondary component of tunneling injected current density for various surface mobility ratio. The surface mobility ratio is defined as the ratio of surface mobility to bulk mobility

calculations suggest that a higher response speed is realized by decreasing the surface mobility ratio from 0.0001 to 0.1, that is, the value of the surface mobility approaches that of the bulk mobility, as shown in Fig. 7.5b. Thus, we expected that a decrease in the number of molecular blind alleys would be effective in obtaining a higher speed of response, despite the lower multiplication rate expected simultaneously.

A more detailed study is needed to clarify the relationships between the response speed and the characteristics of molecular blind alleys, such as spatial depth and areal density.

Numerical calculations electromagnetically justified the sufficient hole accumulation to cause photocurrent multiplication based on the structural trap model with several tens of nanometer-sized metal nanoparticles and molecular blind alleys (Figs. 7.2. and 7.3). This can be regarded as strong support for the proposed structural trap model.

7.2.5 *Keys of High-Speed Response*

Based on the transient photocurrent behavior and numerical calculations, we considered the following to be effective for fabricating high-speed response devices.

(i) **Increase of the quantum efficiency of primary photocurrent**

The quantum efficiency of the primary photocurrent was only 0.65% for the ITO(indium tin oxide)/Me-PTC/Au cell (Fig. 7.1a, red circles). By increasing this quantum efficiency, a shorter onset time of the multiplied photocurrent (2nd component) (Fig. 7.1a, blue circle) would be obtained, and as a result, a high response speed would be observed. For a quantum efficiency of the primary photocurrent close to 100%, an approximately 100 times faster response might be possible compared to the ITO/Me-PTC/Au cell.

(ii) **Control of the characters of molecular blind alleys**
- **Smaller number of molecular blind alleys**

The response speed corresponds to the time required to reach a steady state of the number of holes captured by the molecular blind alleys. Numerical calculations suggest that the time required to establish a steady state is shortened for a smaller surface mobility ratio. Thus, a higher response speed is expected for a smaller number of molecular blind alleys, although a lower multiplication rate is expected simultaneously.

Based on these keys, two types of high-speed multiplication devices were demonstrated.

7.3 High-Speed Response Devices Having C_{60}:CuPc Co-deposited Films

High-speed response devices with a co-deposited film of fullerene and copper phthalocyanine (C_{60}:CuPc) were fabricated. Rise and decay times of 8 and 15 ms were observed for light-on and light-off, respectively, at a multiplication rate of 15-fold.

7.3.1 Motivation

In the field of organic solar cells, a photocarrier generation efficiency of approximately 80% has been achieved using donor–acceptor sensitization in co-deposited film [7–11]. Therefore, to obtain a high-speed response by utilizing the large quantum efficiency of the primary photocurrent, multiplication-type devices incorporating donor/acceptor co-deposited films were fabricated [12, 13].

7.3.2 Cell Structure

We chose the combination of C_{60} acting as an acceptor and CuPc acting as a donor to fabricate high-speed response devices because photovoltaic cells utilizing the C_{60}:CuPc co-deposited film showed a photocurrent quantum efficiency of 80% [10, 11]. Sandwich-type cells (Fig. 7.6) were fabricated by vacuum evaporation at a pressure of 10^{-3} Pa. C_{60} and CuPc were co-deposited from two separate controlled sources on an indium tin oxide (ITO) glass substrate [13]. The ratio of C_{60} and CuPc in the co-deposited film was maintained at 7:3, where the maximum quantum efficiency of the primary photocurrent was observed. The thickness of the organic films was 500 nm. A semitransparent Au electrode was then deposited on the organic film. The multiplication behaviors of a single C_{60} film and C_{60}:CuPc co-deposited film under 10^{-1} Pa at room temperature were compared.

7.3.3 Photoresponse of Multiplied Photocurrent

Figure 7.7 shows the photoresponse of the multiplied photocurrent for a single C_{60} cell. The multiplication rate was tenfold. Rise and decay times of 1 and 13 s were observed for light-on and light-off, respectively [13]. Here, the rise and decay times are defined as the time required to reach 90% of the maximum value of the multiplied photocurrent and the time required to return to 90% of the initial value, respectively, before light irradiation.

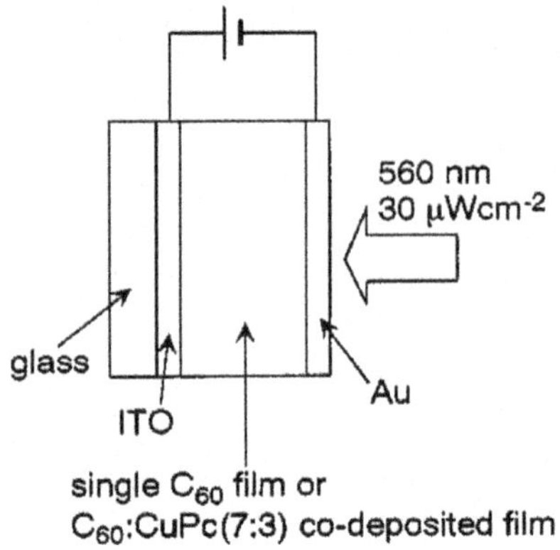

Fig. 7.6 Structure of the sandwich-type cell incorporating a C_{60}:CuPc co-deposited film. Reproduced with the permission of [12]. © 2002 American Institute of Physics

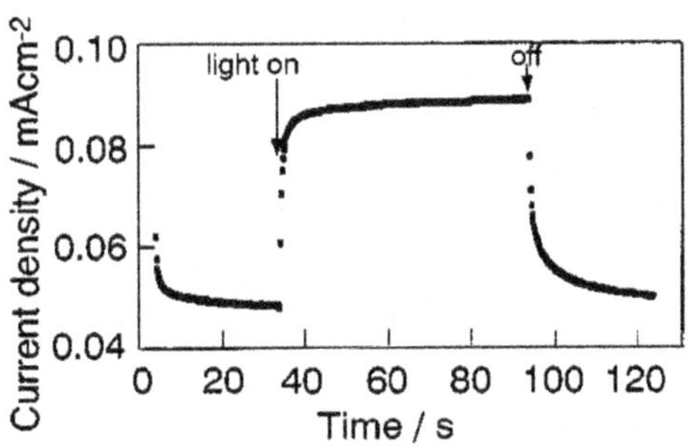

Fig. 7.7 Multiplied photocurrent profile for a single C_{60} cell. Au electrode was biased negatively with respect to the ITO electrode. The dot interval is 0.2 s. The applied voltage is 10 V. Monochromatic light of 580 nm (30 μWcm^{-2}) was irradiated to the Au electrode. Reproduced with the permission of [12]. © 2002 American Institute of Physics

On the contrary, a cell incorporating the C_{60}:CuPc (7:3) co-deposited film, both the rise and decay were completed within the 0.2 s time resolution for Fig. 7.7 under the same measuring conditions. Figure 7.8 (curve a) shows the response profile of the 100 μs time resolution for the same cell under the irradiation of cycled light pulses of 100 ms width (Fig. 7.8, curve b). The multiplication rate was 15-fold. Rise and decay times of 8 and 15 ms were observed for light on and off, respectively. Therefore, this device can operate under light pulses greater than 40 Hz. The response profiles were reproducible for each light pulse and quickly returned to their initial values before irradiation. Since the onset of the multiplied photocurrent completely followed that of the LED, the multiplied photocurrent obviously started to rise within a dot interval of 100 ms. Thus, the photoresponse time was shortened to one-thousandth without lowering the multiplication rate by the addition of CuPc.

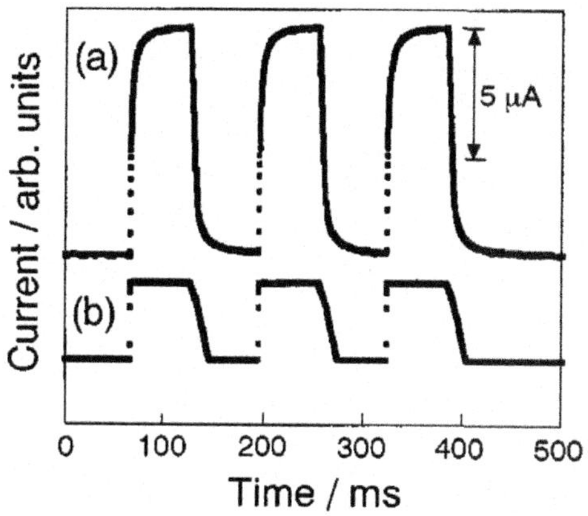

Fig. 7.8 **a** Multiplied photocurrent profile for a cell using C_{60}:CuPc (7:3) co-deposited film. Au electrode was biased negatively with respect to the ITO electrode. The applied voltage is 14 V. Response profiles were measured by using a current amplifier combined with a digital storage oscilloscope. A light emitting diode (LED) light (560 nm, 250 μWcm^{-2}) driven by the step voltage from a function generator was irradiated to Au electrode. **b** The shape of cycled light pulses. The dot interval is 100 μs. Reproduced with the permission of [12]. © 2002 American Institute of Physics

Figure 7.9 shows the dependence of the rise time and multiplication rate on the applied voltage. The rise time was shortened with the applied voltage despite an increase in the multiplication rate. The fastest rise time (3 ms) was observed at 20 V.

Figure 7.10 shows the dependence of the rise time and the multiplied photocurrent on the incident light intensity. A decrease in the light intensity caused little change in the rise time. In addition, the multiplied photocurrent was almost proportional to the light intensity from 15 to 250 mWcm^{-2}. The proportional behavior between the multiplied photocurrent and light intensity is an indispensable characteristic of the photosensor.

Fig. 7.9 The dependence of the rise time (open circles) and multiplication rate (closed squares) on applied voltage for C_{60}:CuPc (7:3) cell. A monochromatic light of 560 nm (250 μW cm^{-2}) is irradiated on Au electrode biased negatively. Reproduced with the permission of [12]. © 2002 American Institute of Physics

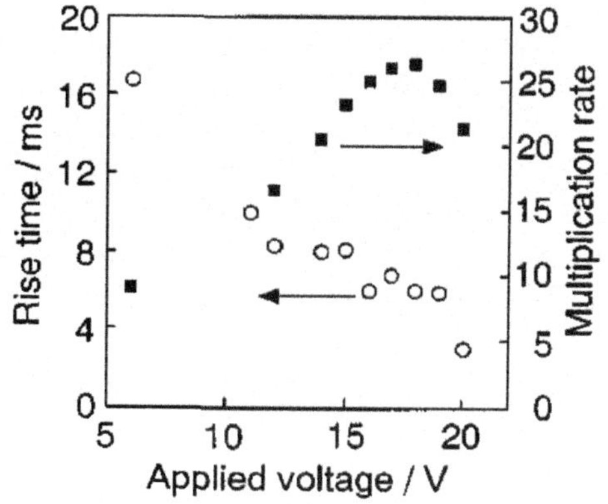

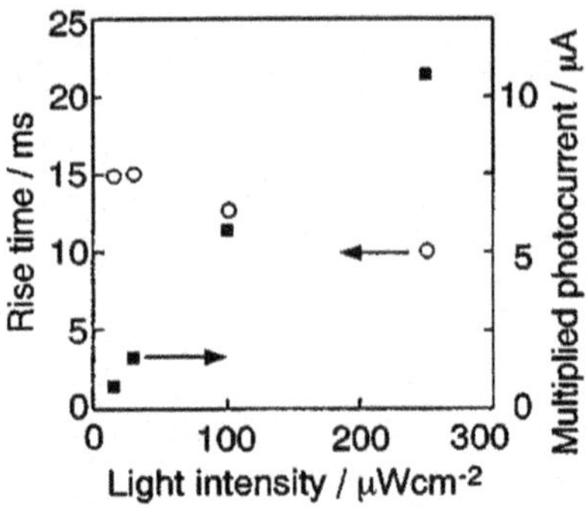

Fig. 7.10 The dependence of the rise time (open circles) and multiplied photocurrent (closed squares) on incident light intensity. The applied voltage is 11 V. A monochromatic light of 560 nm is irradiated on Au electrode biased negatively. Reproduced with the permission of [12]. © 2002 American Institute of Physics

7.3.4 Factors of High-Speed Response

The higher speed response can be attributed to the following two factors occurring in C_{60}:CuPc co-deposited films.

(i) **Increase of the carrier photogeneration efficiency**

The quantum efficiency of the primary photocurrent of the C_{60}:CuPc (7:3) film was measured to be 60%, even at a low voltage of 6 V. Thus, it is supposed to be close to the upper limit of 100% in the high-voltage region. Compared to the ITO/Me-PTC/Au cell with a quantum efficiency of 0.65% (Fig. 7.2a), more than 100 photogenerated holes were supplied to the C_{60}:CuPc/Au interface. Consequently, a sufficient number of holes for electron injection quickly accumulated at the C_{60}:CuPc/Au interface within 10 ms.

(ii) **Smaller number of molecular blind alleys**

Amorphous films showed little multiplication, as described in Chaps. 4 and 5 [6]. Co-deposited films have a strong tendency to form amorphous films because the two types of molecules are mixed at the molecular level [10, 14, 15]. In the case of C_{60}:CuPc (7:3), crystalline CuPc parts with molecular blind alleys decreased. Thus, the time required to establish the steady state is shortened, and a higher response speed is observed under the lower multiplication rate of 15-fold, in accordance with the numerical calculation (Fig. 7.5b).

7.4 High-Speed Devices Having Double-Layered Structure

The double-layered device, composed of polycrystalline NTCDA (Fig. 7.11a) and amorphous perylene derivative (PhEt-PTC) (Fig. 7.11b), showed a high-speed response of 26 ms at a multiplication rate of 1,000-fold [16].

7.4.1 Motivation

A high-speed response was expected by increasing the quantum efficiency of the primary photocurrent. High quantum efficiency of the primary photocurrent was realized under a high applied voltage. For crystalline NTCDA films, a high voltage was not applied because of breakdown. We expected that by inserting a uniform and amorphous organic film, the high-voltage operation of crystalline NTCDA cells, accompanied by a high-speed response, would be possible.

7.4.2 Double-Layered Structure

The double-layered device (Fig. 7.11b) was composed of naphthalene tetracarboxylic anhydride (NTCDA) (400 nm) and phenethyl perylene pigment [*N,N'*-*bis*(phenylethyl)-perylene-3,4:9,10-*bis*(dicarboximide)] (PhEt-PTC) (100 nm). NTCDA acts as the multiplication layer [5, 17], and PhEt-PTC acts as a uniform amorphous layer [6]. For comparison, an NTCDA single-layer device was fabricated (Fig. 7.11a). The total thickness of the organic layers in these devices was the same (500 nm). Photocurrent multiplication occurred at the NTCDA/Au interface biased negatively for both devices. Organic films were prepared on an ITO glass substrate by vacuum evaporation at a rate of 0.2 nms^{-1} under 1×10^{-3} Pa.

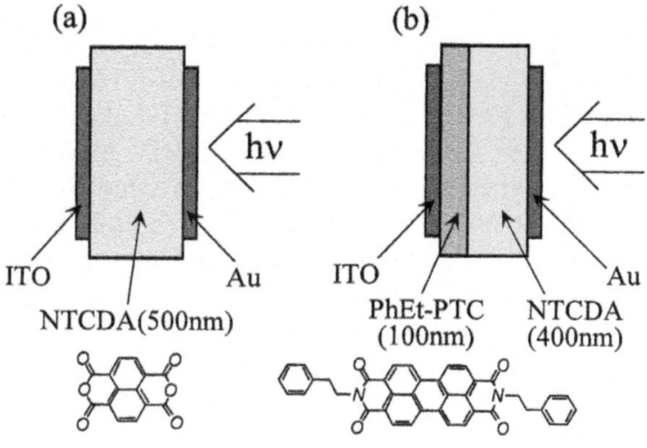

Fig. 7.11 Device structures of the single-layered device (**a**) and the double layered device (**b**). Chemical formulas of NTCDA and Me-PTC are also shown. Reproduced with the permission of [16]. © 2000 American Institute of Physics

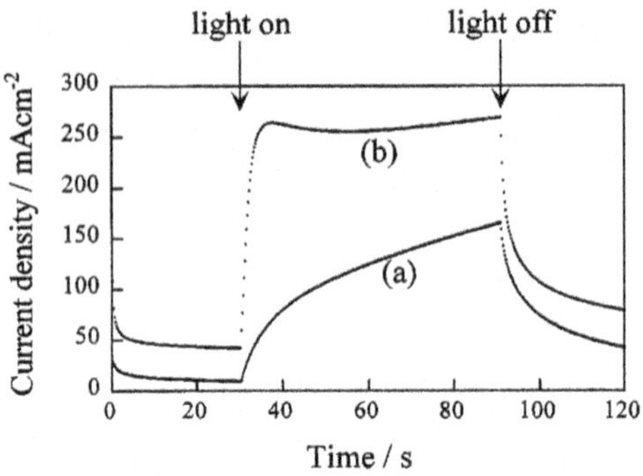

Fig. 7.12 Transient responses of the multiplied photocurrent in **a** a single layered device and **b** a double-layered device. Applied voltages were 12 V (curve a) and 20 V (curve b). Measurements were performed at room temperature. Reproduced with the permission of [16]. © 2000 American Institute of Physics

7.4.3 Transient Response

Figure 7.12 shows typical transient responses for a conventional single-layered cell (curve a) and for a double-layered cell (curve b). In the single-layered cell, the multiplied photocurrent showed a very slow response, and saturation was hardly observed even after prolonged irradiation for 60 s at 12 V. However, in the double-layered cell, a higher operational voltage of 20 V was applied. Thus, the multiplied photocurrent reached a constant value at a multiplication rate exceeding 1.5×10^5-fold after only 4 s of light irradiation. From the initial rise profile measured using an oscilloscope, the multiplication rate was found to reach 470-fold after 10 ms of irradiation, and the time required to reach 1,000-fold was 26 ms.

7.4.4 Film Morphology

Figure 7.13 shows AFM images of (a) NTCDA and (b) PhEt-PTC films deposited on an ITO glass substrate. In the case of NTCDA films, many grains with a size of more than 100 nm were grown on an ITO glass substrate, and the film surface was very uneven. In the case of the PhEt-PTC films, the grain size was much smaller and the surface was relatively uniform. This difference is attributed to the type of substituent. NTCDA molecules with small substituents tend to stack and form large grains. The PhEt-PTC molecules have bulky substituents and hardly form molecular stacks, resulting in small grains and an even surface. The PhEt-PTC film with no X-ray diffraction peaks was amorphous. The double-layered film of these two pigments can be regarded as more uniform than a single NTCDA film.

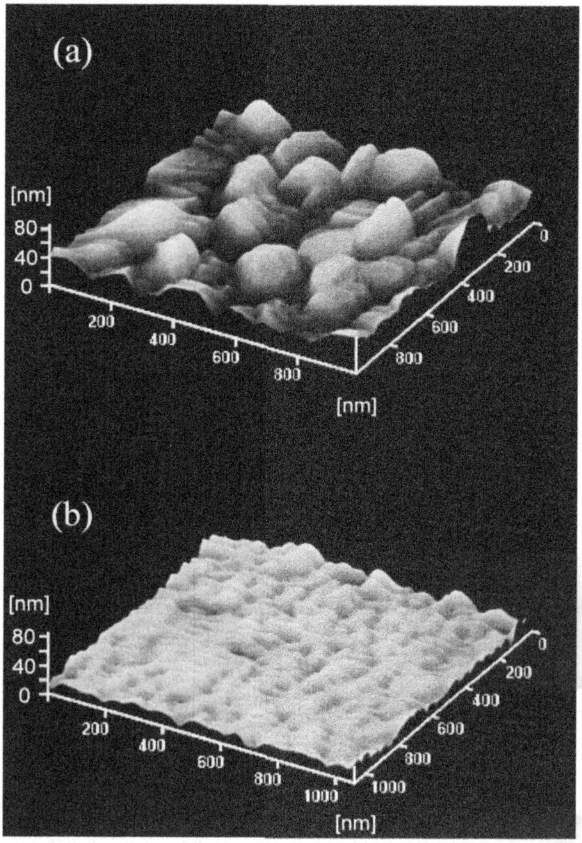

Fig. 7.13 The AFM images of organic films deposited on an ITO glass substrate. **a** NTCDA film (400 nm). **b** PhEt-PTC film (100 nm). Reproduced with the permission of [16]. © 2000 American Institute of Physics

7.4.5 Dark Current Suppression

In the NTCDA single-layered device, the dark current increased drastically with increasing applied voltage, and reached 2.6 $mAcm^{-2}$ at 18 V. In the double-layered device, the dark current was efficiently suppressed to 0.12 $mAcm^{-2}$ at 18 V. In the latter case, the electric field applied to the device was distributed uniformly throughout the entire active area, and a pinpoint high electric field leading to breakdown (usually observed in the former case) was prevented by the improved film uniformity. Consequently, by inserting a PhEt-PTC layer, the dark current was suppressed, and high-voltage application to the device was possible.

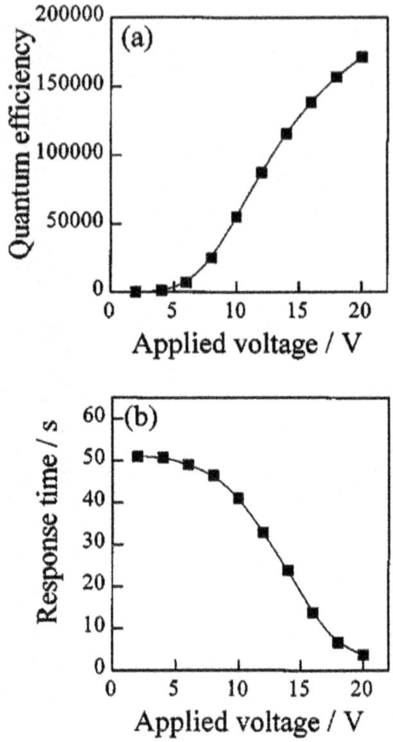

Fig. 7.14 The voltage dependence of **a** multiplication rate and **b** response time of the photocurrent in the double-layered device. The response time of the photocurrent multiplication is defined as the time to reach 90% of the magnitude of the multiplied photocurrent after light irradiation for 60 s. Reproduced with the permission of [16]. © 2000 American Institute of Physics

7.4.6 Multiplication Rate Versus Applied Voltage

Figure 7.14a shows the voltage dependence of the multiplication rate in a double-layered device. As the multiplication rate exceeded 10^5-fold, which is as high as that of the single-layer device [3], the underlying PhEt-PTC layer did not perturb the multiplication occurring at the NTCDA/Au interface. This is because the electrons injected from the Au electrode easily pass through the PhEt-PTC layer and reach the ITO electrode because the conduction band of PhEt-PTC is lower than that of NTCDA [6, 18, 19].

7.4.7 Transient Response Versus Applied Voltage

Figure 7.14b shows the response time of the multiplied photocurrent in the double-layered device. The response time decreased with increasing applied voltage, reaching 3.7 s at 20 V. The transient response began to saturate above 18 V. Although the response time of the single-layered device showed a similar tendency, a voltage higher than 14 V could not be applied owing to breakdown. High-voltage operation in a double-layered device enables high-speed response. Thus, by inserting an amorphous PhEt-PTC layer, a quick response of 3.7 s with a large multiplication rate of 1.7×10^5-fold was achieved.

7.4.8 Factors of High-Speed Response

The higher speed response can be attributed to the following two factors that occur under higher-voltage applications.

(i) **Increase of the quantum efficiency of primary photocurrent**

When the applied voltage was increased, the quantum efficiency of the primary photocurrent increased significantly. Thus, an effective charge supply to structural traps at the organic/metal interface shortens the onset time of tunneling injection during the multiplication process.

(ii) **Total occupation of molecular blind alleys**

There were a specific number of structural traps at the NTCDA/Au interface. When an extremely high electric field is concentrated at the organic/metal interface, holes barely escape from the molecular blind alleys (Fig. 7.2). When most of the molecular blind alleys were occupied, further hole accumulation was limited by the number of molecular blind alleys and ceased. Under these extreme conditions, a high-speed response was achieved.

7.5 Requirements of Organic Multiplication-Type Photosensors

Following requirements should be solved to realize the practical organic multiplication-type photodetectors.

(1) High-speed responses for light-on and-off.

A response speed on the order of milliseconds is necessary for the detection of two-dimensional (2D) images.

(2) Suppression of dark current.

A large signal-to-noise ratio (SNR) is indispensable for the detection of very weak light. Usually, the origin of the dark current is hole injection from the positively biased counter electrode. In the case of multiplication devices, such injected holes can move to the opposite negatively biased organic/metal interface and can also trigger the same multiplication process. Because this process enhances the magnitude of the dark current, the suppression of hole injection from the counter electrode is even more important for multiplication-type devices. A possible solution is to insert a hole-blocking layer at the metal/organic interface that is positively biased [20].

(3) Proportional relationship between the light intensity and the multiplied photocurrent.

For the precise detection of images, the light intensity and magnitude of the multiplied photocurrent should be proportional for a wide range of light intensities.

7.6 Conclusion

We demonstrated high-speed response devices using a C_{60}:CuPc co-deposited layer and an NTCDA/PtEt-PTC double layer. The former device showed a high-speed response of 8 ms for light-on at a multiplication rate of 15-fold. The latter device showed a high-speed response of 26 ms at a multiplication rate of 1,000-fold. To fabricate high-speed response devices, a large quantum efficiency of the primary photocurrent and control of the characteristics of the molecular blind alleys are necessary. A detailed study is needed to clarify the relationship between the characteristics of molecular blind alleys and the response speed. The results in this chapter suggest that organic multiplication-type devices have potential for application in practical imaging devices.

References

1. Nakayama, K., Hiramoto, M., Yokoyama, M.: Direct tracing of the photocurrent multiplication process in an organic pigment film. J. Appl. Phys. **84**, 6154–6156 (1998)
2. Nakayama, K., Hiramoto, M., Yokoyama, M.: Transient response of multiplied photocurrent observed in metal/organic pigment film interface. In: Proceedings of IS&T's NIP14: International Conference on Digital Printing Technologies, Toronto, Ontario, Canada, pp. 490–493, 18–23 Oct 1998
3. Nakayama, K., Hiramoto, M., Yokoyama, M.: Numerical simulations of photocurrent multiplication phenomenon at organic/metal interface. In: Proceedings of IS&T's NIP15: International Conference on Digital Printing Technologies, Orland, Florida, pp. 743–746, 17–22 Oct 1999
4. Nakayama, K., Hiramoto, M., Yokoyama, M.: Numerical simulations of photocurrent multiplication phenomenon process at organic/metal interface. In: Proceedings of Annual Conference of the Imaging Science of Japan, pp. 141–144, Tokyo, 21–23 July 1999. (in Japnaese)
5. Hiramoto, M.: Molecular-sized structural trap at organic-metal interface and photocurrent multiplication phenomenon. In: Salaneck, W.R., Seki, K., Kahn, A., Pireaux, J.-J. (eds.) Conjugated Polymer and Molecular Interfaces (Chap. 18), pp. 585–612. Marcel Dekker Inc., New York/Basel (2002), and references therein
6. Nakayama, K., Hiramoto, M., Yokoyama, M.: Photocurrent multiplication at organic/metal interface and surface morphology of organic films. J. Appl. Phys. **87**, 3365–3369 (2000)
7. Tang, C.W.: Two-layer organic photovoltaic cell. Appl. Phys. Lett. **48**, 183–185 (1986)
8. Hiramoto, M., Fujiwara, H., Yokoyama, M.: Three-layered organic solar cell with a photoactive interlayer of codeposited pigments. Appl. Phys. Lett. **58**, 1062–1064 (1991)

9. Hiramoto, M., Fujiwara, H., Yokoyama, M.: *p-i-n* like behavior in three-layered organic solar cells having c co-deposited interlayer of pigments. J. Appl. Phys. **72**, 3781–3787 (1992)
10. Suemori, K., Miyata, T., Yokoyama, M., Hiramoto, M.: Three-layered organic solar cells incorporating nanostructure-optimized phthalocyanine: fullerene codeposited interlayer. Appl. Phys. Lett. **86**, 063509 (3 pages) (2005)
11. Uchida, S., Xue, J., Rand, B.P., Forrest, S.R.: Organic small molecule solar cells with a homogeneously mixed copper phthalocyanine:C_{60} active layer. Appl. Phys. Lett. **84**, 4218–4220 (2004)
12. Matsunobu, G., Oishi, Y., Yokoyama, M., Hiramoto, M.: High-speed multiplication-type photodetecting device using organic co-deposited films. Appl. Phys. Lett. **81**, 1321–1322 (2002)
13. Hiramoto, M., Yokoyama, M.: High-speed photocurrent multiplication device. Japanese Patent 3426211 (2003)
14. Suemori, K., Miyata, T., Hiramoto, M., Yokoyama, M.: Enhanced photovoltaic performance in fullerene:phthalocyanine co-deposited films deposited on heated substrate. Jpn. J. Appl. Phys. **43**, L1014–L1016 (2004)
15. Suemori, K., Miyata, T., Oishi, Y., Hiramoto, M., Yokoyama, M.: Crystalline-amorphous organic co-deposited films showing efficient photo-electric conversion. Mol. Cryst. Liq. Cryst. **425**, 459–466 (2004)
16. Nakayama, K., Hiramoto, M., Yokoyama, M.: A high-speed photocurrent multiplication device based on an organic double-layered structure. Appl. Phys. Lett. **76**, 1194–1196 (2000)
17. Katsume, T., Hiramoto, M., Yokoyama, M.: Photocurrent multiplication in naphthalene tetracarboxylic anhydride film at room temperature. Appl. Phys. Lett. **69**, 3722–3724 (1996)
18. Kikuchi, M., Takagi, K., Naito, H., Hiramoto, M.: Single crystal organic photovoltaic cells using lateral electron transport. Org. Electron. **41**, 118–121 (2017)
19. Hiramoto, M., Kubo, M., Shinmura, Y., Ishiyama, N., Kaji, T., Sakai, K., Ohno, T., Izaki, M.: Bandgap science for organic solar cells. Electronics **3**, 351–380 (2014)
20. Guo, D., Yang, D., Zhao, J., Vadim, A., Ma, D.: Role of interfaces in controlling charge accumulation and injection in the photodetection performance of photomultiplication-type organic photodetectors. J. Mater. Chem. C (2020). https://doi.org/10.1039/d0tc01590c

Chapter 8
Effect of Oxygen and Water on Photocurrent Multiplication Rates

8.1 Background

Photocurrent multiplication of up to 10^5-fold has been demonstrated in vacuum-evaporated films of both *n*-type (perylene pigment (Me-PTC) [1] and naphthalene tetracarboxylic anhydride (NTCDA) [2]) and *p*-type (quinacridone (DQ)) organic semiconductors (OSCs) [3] (Chap. 2). Photocurrent multiplication in *n*-type OSCs is based on the mechanism of photoinduced electron injection, involving the tunneling injection of electrons from a metal electrode to the organic layer, which is triggered by photogenerated holes at the organic/metal interface [1, 2]. Photocurrent multiplication for *p*-type OSCs is based on the mechanism of photo-induced hole injection, involving the tunneling injection of holes from a metal electrode to the organic layer, which is triggered by photo accumulated electrons at the organic/metal interface [3].

The photoelectrical properties of OSCs are completely different in ultra-high vacuum (UHV) when compared with that in air, owing to the absence of oxygen and water molecules. For example, phthalocyanines, which are *p*-type in air, are *n*-type in UHV [4, 5]. The phenomenon of photocurrent multiplication is no exception.

8.2 Motivation

Photocurrent multiplication for OSC films was found to be sensitive to atmosphere. Hence, we investigated the influence of O_2 on the photocurrent multiplication characteristics of *n*- and *p*-type OSC films. In addition, H_2O molecules influences the photocurrent multiplication. To understand the effect of water molecules alone, in-situ measurements were performed using an ultrahigh vacuum (UHV) apparatus that eliminates water molecules.

M. Hiramoto, *Organic Photocurrent Multiplication*, Electronic Materials: Science & Technology, https://doi.org/10.1007/978-981-99-1237-7_8

8.3 Effect of O_2 Under Ex-Situ Conditions

Multiplication enhancement and suppression by O_2 have been observed in *p*- and *n*-type OSCs, respectively [6].

8.3.1 Ex-Situ Conditions

The measurements presented in this section were performed under the ex-situ condition. The fabricated cells were exposed to ambient atmosphere for several hours prior to the measurement. This implies that H_2O molecules from the ambient atmosphere were adsorbed onto the surface of the OSCs, as described in detail in Sect. 8.4. Multiplication rates in cryostats evacuated to 10^{-1} Pa and filled with O_2 gas (1 atm) at room temperature were compared.

8.3.2 p- *and* n-*Type OSC Cells*

Sandwich-type cells with the same structure, indium tin oxide (ITO)/OSC/Au, were fabricated for *p*-DQ and *n*-NTCDA (Fig. 8.1). Multiplication occurred at the positively biased *p*-DQ/Au junction and the negatively biased *n*-NTCDA/Au junction. To perform a precise comparison between *p*- and *n*-type OSCs, the same metal electrode (Au) was used for the organic/metal junction during the tunneling injection of holes and electrons for *p*-DQ and *n*-NTCDA, respectively.

NTCDA (Fig. 8.1) and DQ (Fig. 8.1) were purified twice using train sublimation [7]. The OSCs and Au were deposited using vacuum evaporation onto ITO glass substrates at a pressure of 10^{-3} Pa.

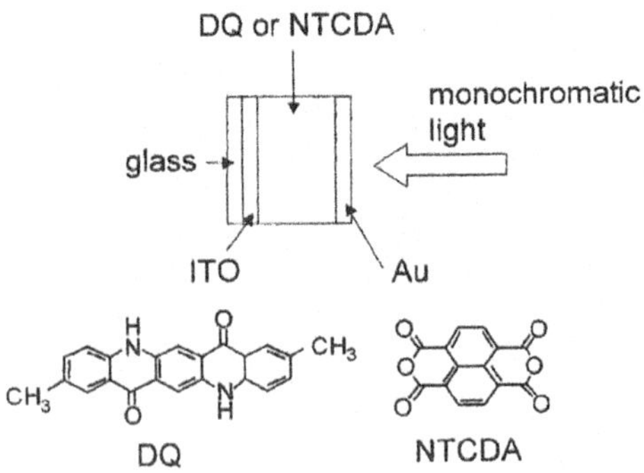

Fig. 8.1 Structure of sandwich-type cells for ex-situ measurements. Thicknesses of DQ, NTCDA, and Au films were 600, 500, and 20 nm. Reproduced with the permission of [6]. © 2003 The Japan Society of Applied Physics

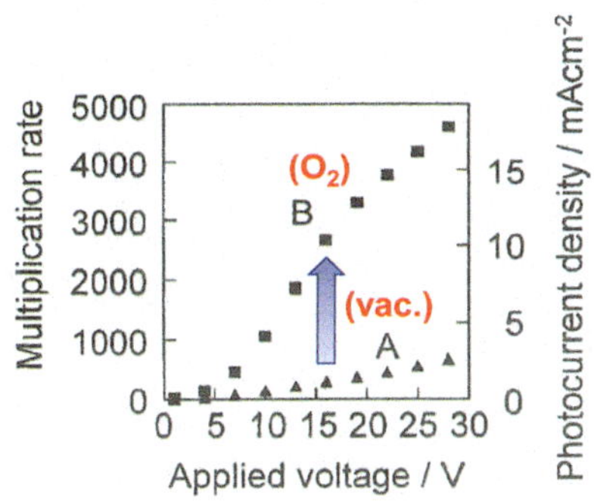

Fig. 8.2 Dependence of the multiplication rate and multiplied photocurrent density on the applied voltage for an ITO/DQ (600 nm)/Au cell (i) at 10^{-1} Pa (curve A) or (ii) under O_2 (1 atm) (curve B). Au electrode was positively biased with respect to the ITO electrode. Monochromatic light of 600 nm (0.035 mW cm^{-2}) was irradiated onto the Au electrode. Measurements were performed at room temperature. Photocurrent quantum efficiency, i.e., multiplication rate, was calculated as the ratio of the number of carriers collected to the number of photons absorbed by the organic film. Reproduced with the permission of [6]. © 2003 The Japan Society of Applied Physics

8.3.3 p-*DQ Cells*

Figure 8.2 shows the dependence of the multiplication rate on the applied voltage for a *p*-type DQ cell. The Au electrode was positively biased with respect to the ITO electrode. Multiplication occurs at the *p*-DQ/Au interface. Although the multiplication rate was only approximately 700-fold at 28 V under vacuum (curve A), it increased to 4800-fold under O_2 atmosphere (curve B).

Figure 8.3 shows the response profile of the multiplied photocurrent for O_2 introduction and evacuation. The magnitude of change in the multiplied photocurrent exceeded 1 mA cm^{-2}. A similar enhancement by oxygen was observed for photocurrent multiplication in *p*-type phthalocyanine cells [8]. Therefore, we conclude that oxygen strongly enhances the photocurrent multiplication of *p*-type OSCs.

8.3.4 n-*NTCDA Cells*

Figure 8.4 shows the dependence of the multiplication rate on the applied voltage for an *n*-type NTCDA cell. The Au electrode was negatively biased with respect to the ITO electrode. Multiplication occurred at the *p*-DQ/Au interface. Although the multiplication rate exceeded 70,000-fold at 5 V under vacuum (curve A), it decreased to approximately 500-fold at 5 V under an O_2 atmosphere (curve B).

Figure 8.5 shows the response profile of the multiplied photocurrent for O_2 introduction and evacuation. The response was reversible and could be repeated several times. The magnitude of change in the multiplied photocurrent exceeded

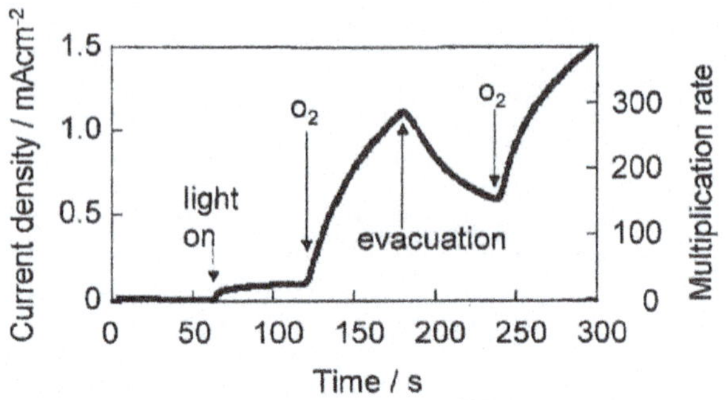

Fig. 8.3 Response profile of the multiplied photocurrent for O_2 introduction and evacuation at 5 V. Au electrode was positively biased with respect to the ITO electrode. Monochromatic light of 600 nm was continuously irradiated after 63 s. Cell was the same as that in Fig. 8.2. Reproduced with the permission of [6]. © 2003 The Japan Society of Applied Physics

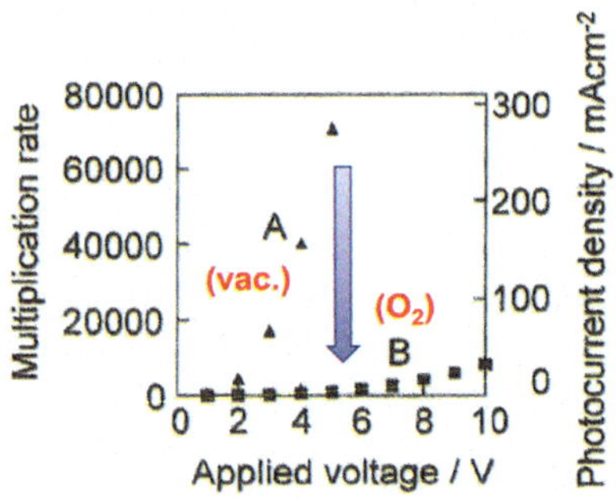

Fig. 8.4 Dependence of the multiplication rate and multiplied photocurrent density on the applied voltage for an ITO/NTCDA (500 nm)/Au cell (i) at 10^{-1} Pa (curve A) or (ii) under O_2 (1 atm) (curve B). Au electrode was negatively biased with respect to the ITO electrode. Monochromatic light of 400 nm (0.066 mW cm^{-2}) was irradiated to the Au electrode. Reproduced with the permission of [6]. © 2003 The Japan Society of Applied Physics

50 mA cm^{-2}. A similar suppression effect by oxygen was also observed for photocurrent multiplication in *n*-type perylene pigment cells (Sect. 8.4) [9]. Therefore, we conclude that oxygen suppresses photocurrent multiplication in *n*-type OSCs.

8.3.5 O_2 Enhancement Mechanism at p-DQ/Au Junction

For *p*-type DQ, multiplication occurs by the tunneling injection of holes because of the accumulation of photogenerated electrons captured by the interfacial structural traps (Chap. 5, Fig. 5.13) near the Au electrode, which causes the electric field to concentrate at the interface (Fig. 8.6a) [3]. On introduction of O_2 molecules, they pass

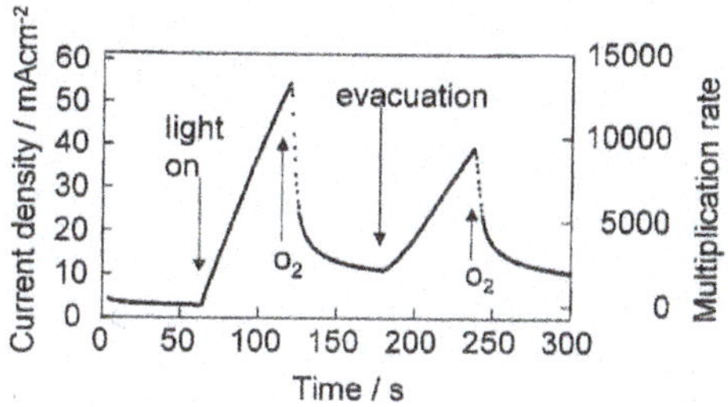

Fig. 8.5 Response profile of the multiplied photocurrent on O_2 introduction and evacuation at 5 V. Au electrode was negatively biased with respect to the ITO electrode. Monochromatic light of 400 nm was continuously irradiated after 63 s. Cell was the same as that in Fig. 8.4. Reproduced with the permission of [6]. © 2003 The Japan Society of Applied Physics

through the 20-nm-thick Au film, which has a gathering structure of nanoparticles with many spatial gaps [10–13]. The O_2 molecules are adsorbed onto the surface of microcrystalline DQ and function as additional electron traps by forming negative oxygen ions (O_2^-) leading to an increase in the multiplication rate (Fig. 8.6b).

Fig. 8.6 Energy structures of the DQ/Au interface biased positively during multiplication in **a** vacuum and **b** O_2 atmosphere. Reproduced with the permission of [6]. © 2003 The Japan Society of Applied Physics

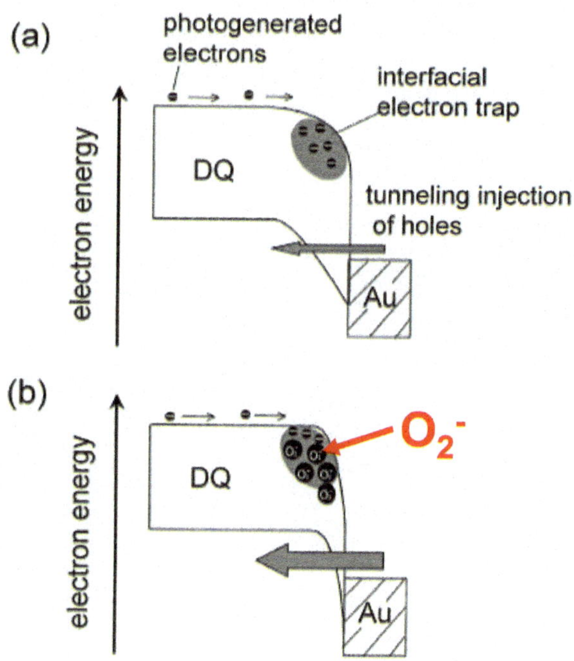

8.3.6 O_2 Suppression Mechanism at n-NTCDA/Au Junction

For *n*-type NTCDA, multiplication occurs by the tunneling injection of electrons owing to the accumulation of photogenerated holes captured by the interfacial traps near the Au electrode, which cause the electric field to concentrate at the interface (Fig. 8.7a) [2]. The presence of O_2 results in reduction in the effective number of positive charges of captured holes at the structural traps and a lowering of the multiplication rate (Chap. 5, Fig. 5.13).

The increase and decrease of multiplication rates by O_2 for *p*-DQ and *n*-NTCDA originate from the tunneling injection of holes and electrons in the multiplication mechanisms between *p*- and *n*-type OSCs. In other words, the present results further support the front-back relationship of the multiplication mechanisms for *p*- and *n*-type OSCs proposed in Chap. 2.

O_2 adsorption is known to induce a change in the dark conductivity of OSCs. Our studies indicate that such a conductivity change occurred mainly in the region below 10^{-4} Pa, and a change in conductivity was rarely observed in the region between 10^{-1} Pa and 1 atm. Further, O_2 adsorption could affect the barrier height for carrier injection. However, we confirmed that such a barrier height change is minimal as is evident for *p*-type phthalocyanine (<0.1 eV) [8]. Therefore, we conclude that the adsorbed O_2 molecules directly affect the multiplication process by acting as electron traps.

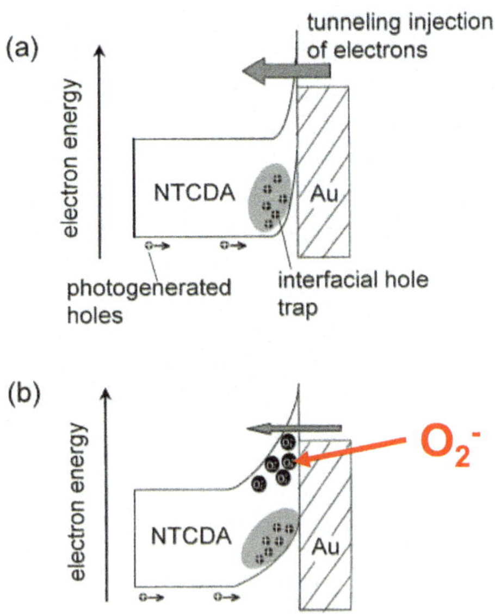

Fig. 8.7 Energy structures of the NTCDA/Au interface biased negatively during multiplication in **a** vacuum and **b** O_2 atmosphere. Reproduced with the permission of [6]. © 2003 The Japan Society of Applied Physics

8.3.7 A Concept of Multiplied Gas Detection

Through the multiplication process, the number of trapped charges is amplified by more than 40,000 times and is observed as the multiplied photocurrent. We observed a significant change in the multiplied photocurrent reaching 270 mA cm^{-2} upon O_2 adsorption (see Fig. 8.4). This indicates that the number of injected carriers reaching 10^{18} s^{-1} cm^{-2} was controlled by O_2 introduction. The change in the number of trapped charges by O_2 introduction was estimated to be of the order of 10^{13} cm^{-2} based on analysis using the Fowler–Nordheim equation [11, 12]. The number of adsorbed gas molecules is amplified by the number of injected carriers for 1 s with the amplification factor reaching 10^5-fold and is detected as the change in the multiplied photocurrent. This suggests that a very small number of gas molecules adsorbed on the surface of the OSCs can be detected by monitoring the multiplied current. The observed effects could be explained by the enhanced detection of adsorbed gas molecules owing to the current multiplication phenomenon.

Notably, the same multiplication process can be triggered by the carriers injected from the counter electrode to induce a multiplied dark current. Therefore, when using the multiplied dark current, light irradiation is not necessary for gas detection.

8.4 Effects of H_2O on Photocurrent Multiplication Observed Under In-Situ Conditions

The adsorbed H_2O molecules strongly enhance photocurrent multiplication for both *p*- and *n*-type OSCs [9].

8.4.1 Irreversible Change Due to H_2O Adsorption

Under the ex-situ conditions described in Sect. 8.3, the fabricated cells were exposed to the ambient atmosphere before the measurements. If adsorbed on the surface of the OSCs, H_2O molecules from the ambient atmosphere are difficult to remove completely, even via vacuum evacuation. Specifically, once the cells fabricated under UHV were exposed to the moisture in the ambient atmosphere, the multiplication rates were changed and did not match the initial multiplication rates even when the cells were returned to UHV.

8.4.2 Motivation

The true nature of the photocurrent multiplication at the OSC/metal interface is obscured by the adsorption of water molecules. This is extremely important from the perspective of the fundamental science of OSCs because it suggests that the real nature of OSCs without adsorbed gases is completely different in the presence of adsorbed gases. Therefore, we chose to investigate the effect of H_2O on photocurrent multiplication by performing cell fabrications and measurements in a UHV chamber. This allowed for the removal of H_2O molecules at any stage of the experiments.

8.4.3 In-Situ Conditions

The in-situ conditions were strictly maintained, and the cell fabrications and measurements were conducted without exposure to ambient atmosphere. An organic molecular beam epitaxy (OMBE) apparatus (EpiQuest Co., Ltd.) equipped with a movable metal mask for patterned metal electrode deposition, pins for electrical contact with electrodes, and a quartz window for light irradiation was used. The purities of the O_2 and N_2 gases used were 99.999 and 99.9999%, respectively, and their H_2O contents were less than 1 and 0.5 ppm, respectively. Photocurrent measurements were performed in the same OMBE chamber under a strictly controlled atmosphere in the following order:

(i) UHV (10^{-4} Pa).
(ii) Oxygen (1 atm).
(iii) Nitrogen (1 atm).

After measuring the multiplication behaviors under in-situ conditions, the same cells were exposed to ambient atmosphere (1 atm) to measure the multiplication rates after the adsorption of H_2O. The measurements were performed in a cryostat. The procedure is outlined below.

(iv) Exposure to ambient atmosphere (1 atm) for 4 h (H_2O adsorption).
(v) Application of 10^{-1} Pa (measurements after H_2O adsorption).

8.4.4 Cells

Sandwich-type cells with the same structure, ITO/OSC/Au, were fabricated for *p*-DQ (quinacridone) and *n*-Me-PTC (perylene pigment) (Fig. 8.8). Multiplication occurred when the *p*-DQ/Au junction was positively biased and the *n*-Me-PTC/Au junction was negatively biased. To perform a precise comparison between the *p*- and *n*-type OSCs, the same metal electrode (Au) was used. The DQ and Me-PTC purified twice by train sublimation [7] were used. DQ, Me-PTC, and Au were evaporated onto an ITO glass substrate from Knudsen cells connected to the OMBE chamber.

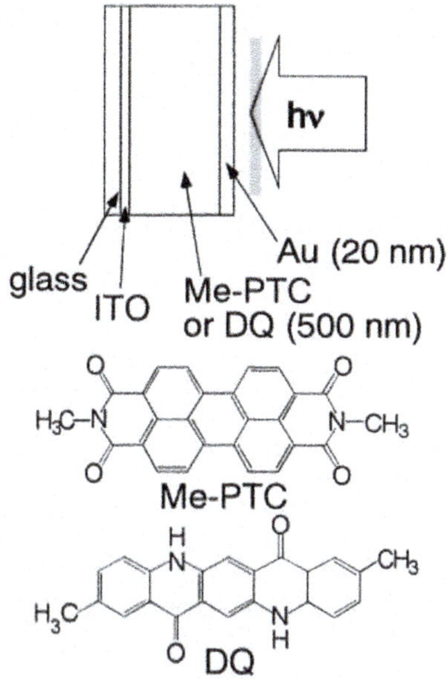

Fig. 8.8 Structure of sandwich-type cells using *n*-Me-PTC and *p*-DQ for in-situ measurements. Reproduced with the permission of [9]. © 2003 The Japan Society of Applied Physics

8.4.5 *Effect of O_2 Under In-Situ Conditions*

Figure 8.9 shows the dependence of the multiplication rates on the applied voltage for *n*-type Me-PTC observed under in-situ conditions in the order of 10^{-4} Pa, O_2, and N_2. Under the 10^{-4} Pa condition, the multiplication rate was approximately 100-fold at 10 V (curve A). Upon O_2 introduction, it was reduced tenfold (curve B). Upon N_2 introduction, the multiplication rate recovered considerably (Curve C).

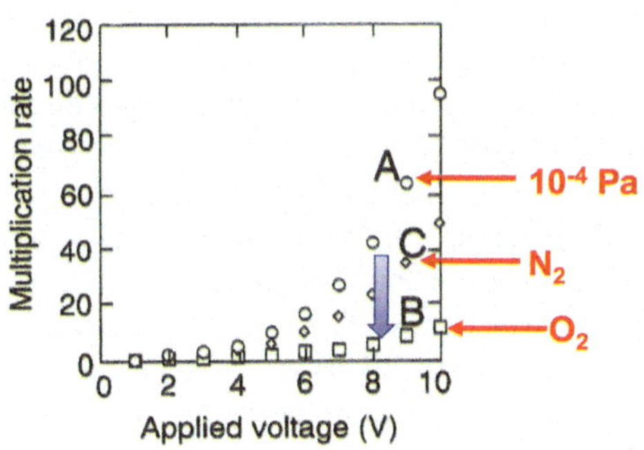

Fig. 8.9 Dependence of the multiplication rate on the applied voltage for an ITO/Me-PTC (500 nm)/Au cell. Measurements were performed at 10^{-4} Pa (curve A), in O_2 (1 atm) (curve B), and in N_2 (1 atm) (curve C). Au was negatively biased with respect to ITO. Monochromatic light (530 nm) obtained from a 500 W xenon lamp through a band pass filter was irradiated on the Au electrode. Reproduced with the permission of [9]. © 2003 The Japan Society of Applied Physics

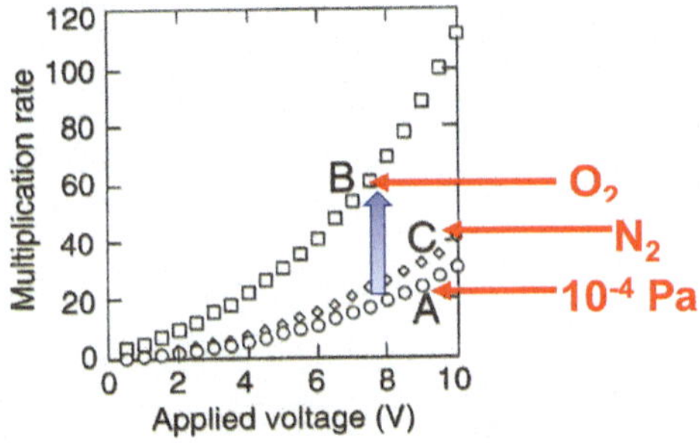

Fig. 8.10 Dependence of the multiplication rate on the applied voltage for an ITO/DQ (500 nm)/Au cell. Measurements were performed at 10^{-4} Pa (curve A), in O_2 (1 atm) (curve B), and in N_2 (1 atm) (curve C). Au was positively biased with respect to ITO. Monochromatic light (590 nm) obtained from a 500 W xenon lamp through a band pass filter was irradiated on the Au electrode. Reproduced with the permission of [9]. © 2003 The Japan Society of Applied Physics

Figure 8.10 shows the results for *p*-type DQ. Under the 10^{-4} Pa condition, the multiplication rate was 30-fold at 10 V (curve A). Upon O_2 introduction, it increased 110-fold (curve B). On N_2 introduction, the rate decreased to its initial value (curve C). The O_2 effects on the multiplication rate showed an opposite trend for the *p*- and *n*-type OSCs. The multiplication rate was suppressed by approximately a factor of 0.1 for *n*-type Me-PTC, while it was enhanced by approximately a factor of 11 for *p*-type DQ. All values of the multiplication rates were lower than those reaching 10^4-fold under ex-situ conditions. Under ex-situ conditions, the adsorbed H_2O molecules remained on the OSC surface [1–3].

8.4.6 *Effect of H_2O*

The multiplication rate was affected significantly on exposure to air. Figures 8.11 and 8.12 show the dependence of the multiplication rate on the applied voltage on exposure to air (curves A), under O_2 (1 atm) in situ conditions prior to exposure (curve B) for the ITO/Me-PTC/Au and ITO/DQ/Au cells, respectively. The multiplication rate at 10 V increased from 10 to 8×10^4-fold and from 1.1×10^2 to 1.8×10^3-fold for Me-PTC and DQ, respectively. On exposure to air, the multiplication enhancement factor reached 8000-fold. The observed effects were caused by the adsorption of molecules other than O_2 and is most likely due to atmospheric moisture. Therefore, we concluded that the marked increase in the multiplication rates observed for both *n*- and *p*-type OSCs were caused by H_2O adsorption.

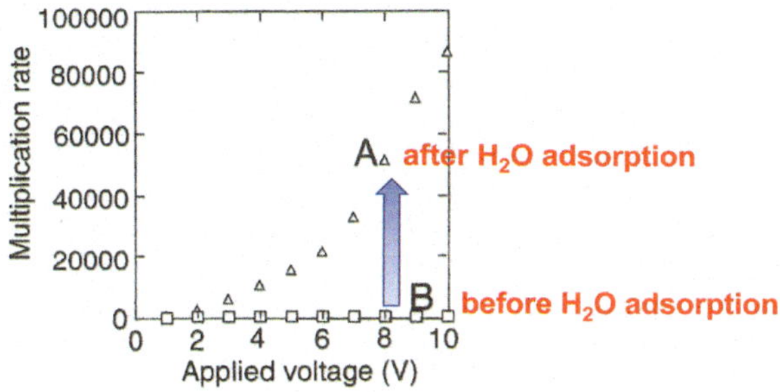

Fig. 8.11 Dependence of the multiplication rate on the applied voltage for an ITO/Me-PTC (500 nm)/Au cell. Measurements were performed on exposure to air (H_2O adsorption) (curve A) and under O_2 (1 atm) in the OMBE chamber prior to exposure to air (H_2O adsorption) (curve B). Au was negatively biased with respect to ITO. Monochromatic light of 530 nm was irradiated on the Au electrode. Reproduced with the permission of [9]. © 2003 The Japan Society of Applied Physics

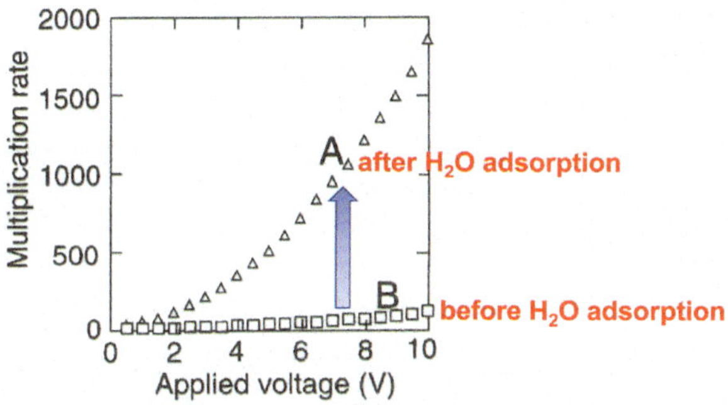

Fig. 8.12 Dependence of the multiplication rate on the applied voltage for an ITO/DQ (500 nm)/Au cell. Measurements were performed on exposure to air (H_2O adsorption) (curve A) and under O_2 (1 atm) in the OMBE chamber prior to exposure to air (H_2O adsorption) (curve B). Au was positively biased with respect to ITO. Monochromatic light of 590 nm was irradiated on the Au electrode. Reproduced with the permission of [9]. © 2003 The Japan Society of Applied Physics

8.4.7 Enhancement/Suppression Mechanisms of O_2 Under In-Situ Conditions

The enhancement/suppression of photocurrent multiplication rates in the presence of O_2 under in-situ conditions were similar to those observed under ex-situ conditions (Sect. 8.3), although the O_2 effects on the multiplication rates were far smaller. Therefore, we conclude that the O_2 effects on the multiplication rate for *p-DQ* and *n*-Me-PTC can be explained based on the same enhancement/suppression mechanisms in Figs. 8.6 and 8.7, respectively.

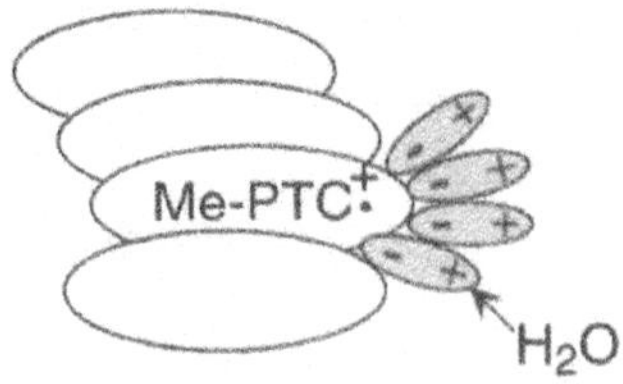

Fig. 8.13 Schematic illustration of the stabilization of Me-PTC.$^+$ molecules owing to hydration. Reproduced with the permission of [9]. © 2003 The Japan Society of Applied Physics

8.4.8 Multiplication Rate Enhancement Mechanism of H_2O

There are two probable causes for the effect of H_2O on multiplication rates.

(i) **Energetic relaxation of trapped charges by the dipoles of adsorbed H_2O molecules**

A trapped hole on the surface can be regarded as a cation radical of the Me-PTC molecule, that is, Me-PTC.$^+$. The polar H_2O molecules adsorbed on the surface can hydrate them (Fig. 8.13). This energy relaxation prevents the release of holes from the molecular blind alleys (Chap. 5, Fig. 5.13) to the Au electrode through the surface; as a result, the multiplication rate is enhanced. This explanation can be also applied to an anion radical of DQ, i.e., DQ.$^-$, and to that of adsorbed O_2^- negative ions.

(ii) **Sensitization of primary photocarrier generation by H_2O molecules**

For ITO/Me-PTC/Au cells, a small photocurrent response was observed below 1 V at 10^{-4} Pa, although short-circuit photocurrent owing to the photovoltaic effect was observed in air [14]. The primary photocarrier generation ability of the Me-PTC was considerably reduced under high vacuum. Titanyl phthalocyanine and H_2O adsorption have been reported to significantly promote photocarrier generation [15, 16]. These findings suggest that the primary photocarrier generation efficiency of OSCs is promoted by H_2O molecules. This phenomenon could lead to an increase in the number of holes at the interface and a multiplication enhancement.

8.4.9 Future Work

(i) **Origin of *p*- and *n*-type behavior under in-situ conditions**

Even under in situ conditions, DQ and Me-PTC exhibited multiplication characteristics, indicating that they behave as *p*- and *n*-type OSCs, respectively. Multiplication enhancement and suppression in the presence of O_2 were observed for DQ and Me-PTC, respectively. The reason for the difference in *p*- and *n*-behaviors under in-situ conditions remains unknown. One possibility is that these *p*- and *n*-type OSCs contain impurities that act as acceptors and donors. Although this is a fundamental issue for OSCs, it is yet to be addressed.

(ii) **Multiplication behavior of doped OSCs**

Doping is crucial for OSCs [17–20]. Recently, complete *pn* control via acceptor and donor doping has been reported [21]. For example, Me-PTC, C_{60}, and phthalocyanines showed both *n*-type and *p*-type behaviors via donor (Cs_2CO_3) and acceptor (MoO_3) doping. Reported literature has indicated the *pn*-control of co-deposited films, such as C_{60}: phthalocyanine [22, 23]. It is important to determine whether the tunneling injections of holes and electrons in multiplication processes can be observed for intentionally doped *p*- and *n*-type OSCs, respectively. Accordingly, the doping technique could allow for the future development of photocurrent multiplication devices using intentionally doped organic/metal junctions, organic heterojunctions, and organic *pn*-homojunctions among others.

8.5 Conclusion

Both in-situ and ex-situ experiments revealed that the adsorbed O_2 molecules suppressed the multiplication rates of *n*-Me-PTC and enhanced the multiplication rates of *p*-DQ. In-situ experiments revealed that the adsorbed H_2O molecules strongly enhanced photocurrent multiplication, irrespective of the conduction type. In conclusion, we propose the concept of enhanced gas detection via photocurrent multiplication [24], which is a novel addition to the literature on organic gas sensors reported thus far [25, 26].

References

1. Hiramoto, M., Imahigashi, T., Yokoyama, M.: Photocurrent multiplication in organic pigment films. Appl. Phys. Lett. **64**, 187–189 (1994)
2. Katsume, T., Hiramoto, M., Yokoyama, M.: Photocurrent multiplication in naphthalene tetracarboxylic anhydride film at room temperature. Appl. Phys. Lett. **69**, 3722–3724 (1996)
3. Hiramoto, M., Kawase, S., Yokoyama, M.: Photoinduced hole injection multiplication in p-type quinacridone pigment films. Jpn. J. Appl. Phys. **35**, L349–L351 (1996)
4. Martin, M., Andre, J.-J., Simon, J.: Influence of dioxygen on the junction properties of metallophthalocyanine based devices. J. Appl. Phys. **54**, 2792–2794 (1983)
5. Tada, H., Touda, H., Takada, M., Matsushige, K.: Quasi-intrinsic semiconducting state of titanyl-phthalocyanine films obtained under ultrahigh vacuum conditions. Appl. Phys. Lett. **76**, 873–875 (2000)
6. Hiramoto, M., Suemori, K., Yokoyama, M.: Influence of oxygen on photocurrent multiplication phenomenon at organic/metal interface. Jpn. J. Appl. Phys. **42**, 2495–2497 (2003)

7. Wagner, H.J., Loutfy, R.O., Hsiao, C.: Purification and characterization of phthalocyanines. J. Mater. Sci. **17**, 2781–2791 (1982)
8. Yoshida, M., Hiramoto, M., Yokoyama, M.: Ext. Abstr. (61st Autumn Meet. 2000). Jpn. Soc. Appl. Phys. Relat. Soc. 3a-L-11 (in Japanese)
9. Hiramoto, M., Fujino, K., Yoshida, M., Yokoyama, M.: Influence of oxygen and water on photocurrent multiplication in organic semiconductor films. Jpn. J. Appl. Phys. **42**, 672–675 (2003)
10. Hiramoto, M., Sato, I., Yokoyama, M.: Metal nanostructure of metal/organic interface causing photocurrent multiplication phenomenon. IEICE Trans. Electron. **E85-C**(6), 1253–1255 (2002)
11. Hiramoto, M.: Molecular-sized structural trap at organic-metal interface and photocurrent multiplication phenomenon. In: Salaneck, W.R., Seki, K., Kahn, A., Pireaux, J.-J. (eds.) Conjugated polymer and molecular interfaces, Chap. 18, pp. 585–612, Marcel Dekker Inc., New York/Basel (2002)
12. Hiramoto, M., Nakayama, K., Katsume, T., Yokoyama, M.: Field-activated structural traps at organic pigment/metal interfaces causing photocurrent multiplication phenomena. Appl. Phys. Lett. **73**, 2627–2629 (1998)
13. Hiramoto, M.: A concept of molecular blind alleys and organic multiplication-type photosensor. In: Fabrication of organic films and application to organic devices, pp. 70–79. CMC (2008) (in Japanese)
14. Hiramoto, M., Ihara, K., Fukusumi, H., Yokoyama, M.: Conduction type control from n- to p-type for organic pigment film purified by reactive sublimation. J. Appl. Phys. **78**, 7153–7157 (1995)
15. Fujimaki, Y.: Proc. IS&T's 7th Int. Congr. Advances in Non-Impact Printing Technologies, vol. 1, p. 269. Portland, Oregon, Oct. 6–11 (1991)
16. Yamaguchi, S., Sasaki, Y.: Effect of water on primary photocarrier-generation process in Y-form titanyl phthalocyanine. J. Phys. Chem. B **104**, 9225–9229 (2000)
17. Tietze, M.L., Pahner, P., Schmidt, K., Leo, K., Lüssem, B.: Doped organic semiconductors: trap-filling, impurity saturation, and reserve regimes. Adv. Funct. Mater. **25**, 2701–2707 (2015)
18. Lüssem, B., Riede, M., Leo, K.: Doping of organic semiconductors. Phys. Status Solidi A **210**, 9–43 (2013)
19. Walzer, K., Maennig, B., Pfeiffer, M., Leo, K.: Highly efficient organic devices based on electrically doped transport layers. Chem. Rev. **107**, 1233–1271 (2007)
20. Salzmann, I., Heimel, G., Oehzelt, M., Winkler, S., Koch, N.: Molecular electrical doping of organic semiconductors: fundamental mechanisms and emerging dopant design rules. Acc. Chem. Res. **49**, 370–378 (2016)
21. Hiramoto, M., Kikuchi, M., Izawa, S.: Parts-per-million-level doping effects in organic semiconductor films and organic single crystals. Adv. Mater. **30**, 1801236 (15 pp) (2018)
22. Ishiyama, N., Kubo, M., Kaji, T., Hiramoto, M.: Doping-based control of the energetic structure of photovoltaic co-deposited films. Appl. Phys. Lett. **99**, 133301 (3 pp) (2011)
23. Kubo, M., Kaji, T., Hiramoto, M.: *pn*-Homojunction organic solar cells formed in the thick phase-separated co-deposited films by doping. Appl. Phys. Lett. **103**, 263303 (2013)
24. Hiramoto, M., Yoshida, M., Yokoyama, M.: Gas detection method and gas sensor utilizing the photocurrent multiplication and so on. Japanese Patent, No. 3530471 (2004)
25. Wu, M., Hou, S., Yu, X., Yu, Y.: Recent progress in chemical gas sensors based on organic thin film transistors. J. Mater. Chem. C **8**, 13482–13500 (2020) and references therein
26. Kumar, A., Meunier-Prest, R., Bouvet, M.: Organic heterojunction devices based on phthalocyanines: a new approach to gas chemosensing. Sensors **20**, 4700 (25 pp) (2020)

Chapter 9
Multiplied Photocurrent Oscillation with Negative Resistance

9.1 Motivation

We have observed large photocurrent multiplication, achieving 10^5-fold, in vacuum-evaporated films of organic heterojunctions, such as CuPc/Me-PTC [1, 2] and Me-PTC/TPD (Chap. 6). This multiplication mechanism is based on electron injection from a negatively biased valence band of CuPc to the conduction band of Me-PTC, which is triggered by hole accumulation at the Me-PTC/CuPc interface. Holes are trapped by molecular blind alleys at the surface of the Me-PTC crystals (Chap. 5). We have also observed photocurrent multiplication at the NTCDA/H_2Pc heterojunction based on the same mechanism, that is, electron injection from a negatively biased valence band of H_2Pc to the conduction band of NTCDA, which is triggered by hole accumulation at the molecular blind alleys at the NTCDA surface (Chap. 5) (Fig. 9.13, lower right) [3, 4].

During the course of our study of photocurrent multiplication at the NTCDA/H_2Pc heterojunction, we accidentally observed photocurrent oscillation under conditions of very large photocurrent multiplication. Because this new phenomenon has not previously been observed for organic semiconductors, the author has proposed investigating the characteristics of this new oscillation phenomenon and the characterization of its mechanism.

9.2 Cell Structure and Measurement Conditions

H_2Pc and NTCDA were purified twice by train sublimation [5], and the Ag counter electrode was deposited by vacuum evaporation on the ITO glass substrate (Fig. 9.2a). The typical thickness of NTCDA was 1 μ m, while that of the H_2Pc films varied from 0 to 60 nm. Four cells were fabricated in the overlapping regions of the four Ag electrodes and a common ITO electrode (Fig. 9.2b).

M. Hiramoto, *Organic Photocurrent Multiplication*, Electronic Materials: Science & Technology, https://doi.org/10.1007/978-981-99-1237-7_9

Voltage- and current-controlled measurements (Fig. 9.3) were performed under irradiation by monochromatic light (400 nm) on the ITO substrate at temperatures ranging from 5 to 288 K, under a pressure of 5×10^{-5} Torr. The ITO was negatively biased with respect to the Ag electrode.

(1) **Voltage-controlled measurements**: The magnitude of the multiplied photocurrent was measured by controlling the applied voltage (Fig. 9.3a).
(2) **Current-controlled measurements**: The magnitude of the photovoltage was measured by controlling the multiplied photocurrent (Fig. 9.3b).

9.3 Photocurrent Oscillation and Voltage-Controlled Measurements

Photocurrent oscillations were first discovered through voltage-controlled measurements at 5 K (Fig. 9.3a). Figure 9.4a shows a typical photocurrent response. At 35 V, a slow rise and decay of several seconds, which had emerged in the context of photocurrent multiplication at the organic heterojunction (Chap. 6), was observed by light irradiation. At a higher applied voltage of 39 V, a remarkable photocurrent oscillation between 0.5 and 1.5 Acm^{-2} appeared. The oscillation amplitude reached a relatively high photocurrent density of approximately 1 Acm^{-2}. This demonstrates such large photocurrent densities are achievable. Figure 9.4c shows the expanded response profile of Fig. 9.4a. Sawtooth waves possessing relatively slow increases and decays were also observed. The oscillation frequency was approximately 1 Hz. Figure 9.4d shows the fast Fourier transform (FFT) analysis of the observed photocurrent oscillation. The spectral peak was located at a frequency of 1.02 Hz.

A similar photocurrent oscillation was observed at room temperature (288 K) (Fig. 9.5a). At 25 V, a smooth photocurrent response was observed. At a higher voltage of 31 V, the photocurrent oscillated between 1.7 and 4 Acm^{-2}. Oscillation amplitude reached above 2.5 Acm^{-2}. Figure 9.4c shows the expanded response profile of Fig. 9.4a. Each wave was characterized by a slow rise and decay of approximately 2 and 1 s, respectively. The FFT analysis (Fig. 9.5c) indicates a spectrum peak at a frequency of 0.3 Hz. Photocurrent oscillation was observed at 77 (Fig. 9.6a), 193 (Fig. 9.6b), and 273 K (Fig. 9.6c). Thus, the photocurrent oscillation phenomenon was found to occur over a wide temperature range (5–288 K).

Figure 9.7 shows the dependence of the multiplication rate on the applied voltage at 5 K. At 34 V, the multiplication rate was 10^4-fold. Above 35 V, the multiplication rate accelerated significantly relative to the extrapolated line (broken line); oscillation between the blue dots commenced at 38 V.

Figure 9.8 shows the absorption spectra of the NTCDA and H_2Pc films. An oscillating photocurrent appeared upon the excitation of NTCDA (400 nm, blue arrow), while no oscillated photocurrent appeared following the excitation of H_2Pc (633 nm, red arrow). NTCDA excitation therefore essential to photocurrent oscillation.

Features of the photocurrent oscillation are summarized as follows.

(i) Maximum amplitude of oscillation reached a relatively large value of 3 Acm^{-2}.
(ii) Oscillation frequency of 0.1–2 Hz.
(iii) Oscillation occurred from very low temperatures (5 K) to room temperature (288 K).
(iv) Oscillation was induced exclusively by the light excitation of NTCDA.

This photocurrent oscillation can be regarded as an inverter function, converting the current from DC to AC.

9.4 S-shaped Negative Resistance: Current-Controlled Measurements

To characterize the origin of the photocurrent oscillation, current-controlled measurements (Fig. 9.3b) were performed. Specifically, the magnitude of the photovoltage was monitored through controlled scanning of the multiplied photocurrent and the excitation of the NTCDA film under 400 nm monochromatic light (225 μWcm^{-2}).

Figures 9.4b and 9.5b show the current density (I)-voltage (V) curves observed through current-controlled measurements at 5 and 288 K, respectively, and for the same device as in Fig. 9.2. The oscillating photocurrents observed through voltage-controlled measurements are also shown in Figs. 9.4a and 9.5a. Between 0.5 and 1.5 Acm^{-2} at 5 K and between 1.7 and 4 Acm^{-2} at 288 K, in spite of the increase in the current density, the corresponding voltage decreased and—clear S-shaped I-V relationships emerged. These observations are consistent with negative resistance phenomena, that is, $dV/dI < 0$.

In the negative resistance region of the S-shaped curves (i.e., $dV/dI < 0$), more than two photocurrent density values may exist at a single voltage. Therefore, the magnitude of the current becomes unstable and oscillates. The combined observation of both the negative resistance of the voltage and current regions associated with the current oscillation is highly suggestive. Thus, we conclude that the photocurrent oscillation was caused by the S-shaped negative resistance.

9.5 Negative Resistance and Heterogeneity

N- and S-type negative resistances are both classes of negative resistance phenomena. Typical examples of the former and latter negative resistances are the Esaki diodes [6–10] and the DC-AC convertor (i.e., thyristors [11]), respectively. For the N-type negative resistance, more than two values of voltage are allowed at a single current value on the I-V curve (Fig. 9.9a). For the S-type negative resistance, more than two current values at a single voltage value are allowed in the I-V curve (Fig. 9.9b) [12, 13]. As a result, the voltage and current become unstable for the N- and S-type negative resistances, respectively, resulting in oscillations.

Figure 9.9c, d present schematics of the heterogeneity of N- and S-type negative resistances, respectively [12]. For the N-type negative resistance (Fig. 9.9a, blue region), low and high electric field (E) regions formed perpendicular to the direction of current flow (pale blue arrow) (Fig. 9.9c). For the S-type negative resistance (Fig. 9.9b, red region), low and high current density (j) regions formed parallel to the direction of current flow (pale blue arrow) (Fig. 9.9d). In the red region in Fig. 9.9b, subtle heterogeneity induces the formation of high- and low-current–density regions that disturb the current stability (Fig. 9.9d) and result in current fluctuation. Conversely, a high current density region decreases the required voltage on the I-V curve of the S-type negative-resistance region (red region).

9.6 Thickness Dependence

9.6.1 H_2Pc Thickness

Multiplication occurs by electron injection from the valence band of H_2Pc to the conduction band of the NTCDA layer, which is triggered by the photogenerated holes at the NTCDA/H_2Pc interface (Fig. 9.1). As shown in Fig. 9.9d, for the S-shaped negative resistance, subtle heterogeneity induces the formation of high and low current density regions and results in current fluctuation, significantly indicating the presence of heterogeneity in the NTCDA/H_2Pc heterojunction. This leads to the expectation that heterogeneity in the H_2Pc thickness may have induced the formation of high and low current density regions, given that the magnitude of the concentrated high electric field across the H_2Pc film would also affect the magnitude of electron injection via the multiplication process (Fig. 9.1).

To validate the above hypothesis, the thickness of the H_2Pc film was varied between 0, 6, 36, 48, and 60 nm while maintaining an NTCDA thickness of 1 μm. Figure 9.10 shows the photocurrent response profiles for various H_2Pc thicknesses. At 0 nm, no photoresponse was observed. At 6 nm, the photoresponse and sign of the

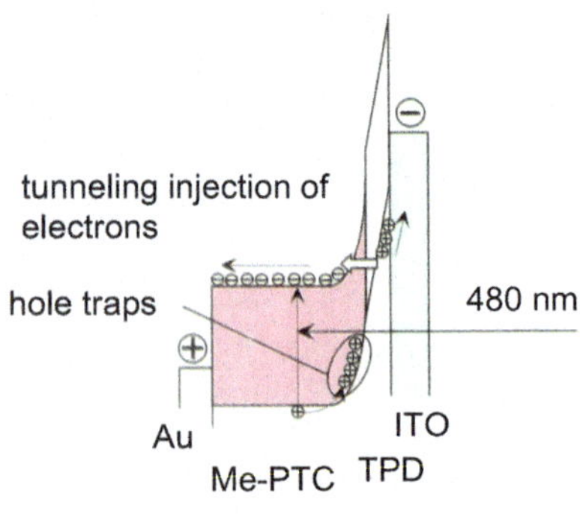

Fig. 9.1 Energy structure during the photocurrent multiplication at Me-PTC/TPD heterojunction

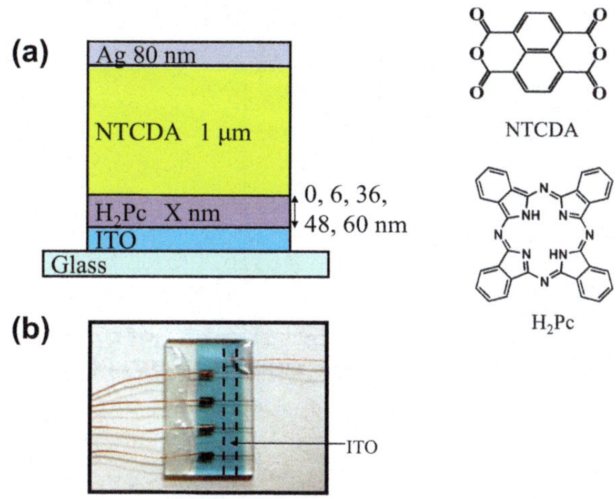

Fig. 9.2 **a** Cell structure of NTCDA/H_2Pc heterojunction sandwiched between ITO and Ag electrodes. The chemical formulas of NTCDA and H_2Pc are also shown. **b** Photograph of cell. The ITO electrode is in between two broken curves

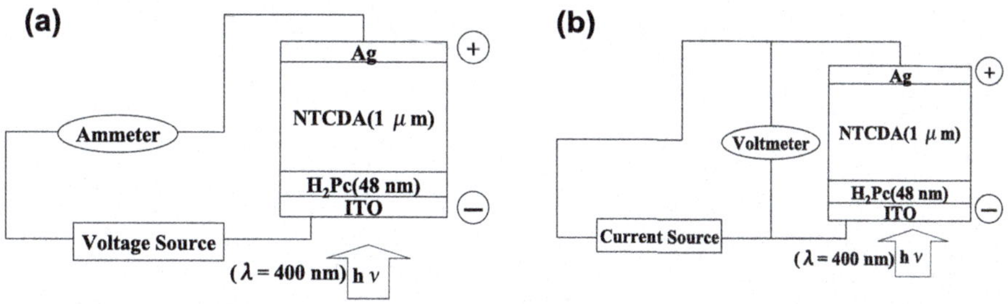

Fig. 9.3 Schematic of **a** voltage-controlled and **b** current-controlled measurements

oscillation could be observed at approximately 40 s. At 36 nm, a photocurrent oscillation possessing an amplitude between 1.5 and 4 Acm^{-2} was observed. At 48 nm, the amplitude of the photocurrent oscillation decreased to between 0.8 and 1.2 Acm^{-2}. At 60 nm, the frequency of the photocurrent oscillation decreased. Evidently, the H_2Pc thickness strongly influences the oscillation behavior.

Figure 9.11a, b show the AFM image and height profile of the H_2Pc film (48 nm) on the ITO substrate, respectively. The H_2Pc film was formed by gathering tiny crystals, each with a diameter of approximately 100 nm. The average roughness was 20 nm. Figure 9.11c illustrates the cross-section of the H_2Pc film reproduced precisely based on Fig. 9.11b. The thickness of the H_2Pc film (48 nm) possessed a lateral spatial heterogeneity of approximately 100 nm and a vertical heterogeneity of approximately 20 nm.

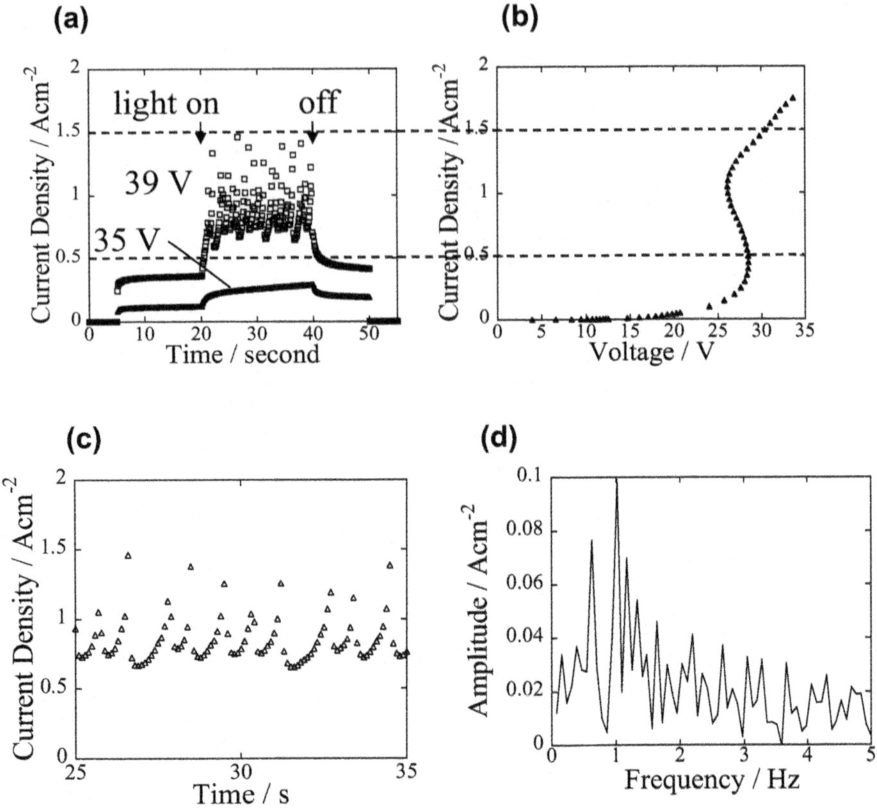

Fig. 9.4 **a** Typical response of the multiplied photocurrent at 5 K by the voltage-controlled measurements. Photocurrent oscillation is observed at 39 V. **b** Current density (I)-voltage (V) curves observed by the current-controlled measurements at 5 K. **c** Expanded response profile of photocurrent oscillation in Fig. 9.4a. **d** Fast Fourier transform (FFT) analysis of the observed photocurrent oscillation in Fig. 9.4a. Thicknesses of the NTCDA and H_2Pc films were 1 μm and 48 nm, respectively. Monochromatic light of 400 nm (225 μWcm^{-2}) was irradiated on ITO electrode. Sampling rate was 10 Hz

9.6.2 NTCDA Thickness

The NTCDA thickness was varied between 0.5, 1.0, 1.5, 2.0, 3.0, and 4.0 μm, while the H_2Pc film thickness was maintained constant (48 nm). Clear photocurrent oscillations were observed at thicknesses other than that of 4.0 μm. Figure 9.12a, b represent the NTCDA thickness dependencies of the onset voltage of photocurrent oscillation and the oscillation region of photocurrent, respectively. Although the NTCDA thickness was increased by a factor of 6 (from 0.5 to 3.0 μm), the onset voltage scarcely changed from an approximate value of 35 V (Fig. 9.12a). The oscillation amplitude between the maximum and minimum oscillation currents was confined between 1 and 2 Acm^{-2} (Fig. 9.12b). Thus, we conclude that the thickness of the NTCDA films has

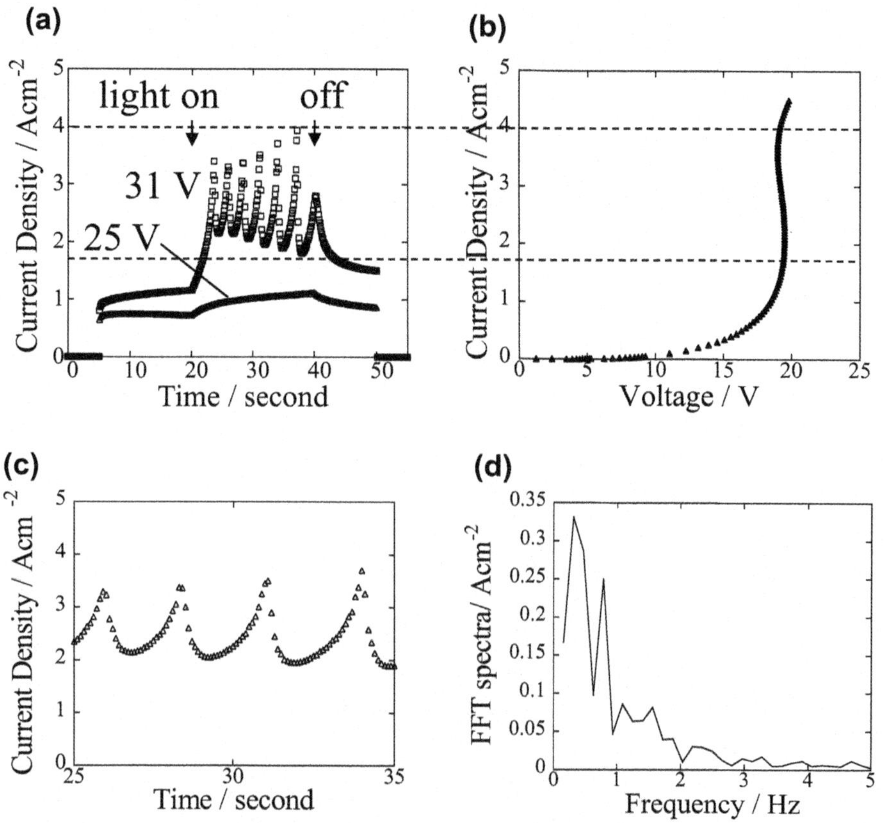

Fig. 9.5 **a** Typical response of the multiplied photocurrent at 288 K by the voltage-controlled measurements. Photocurrent oscillation is appeared at 31 V. **b** Current density (I)-voltage (V) curves observed by the current-controlled measurements at 288 K. **c** Expanded response profile of photocurrent oscillation in Fig. 9.4a. **d** Fast Fourier transform (FFT) analysis of the observed photocurrent oscillation in Fig. 9.5a. Thicknesses of the NTCDA and H_2Pc films were 1 μm and 48 nm, respectively. Monochromatic light of 400 nm (225 μWcm^{-2}) was irradiated on ITO electrode

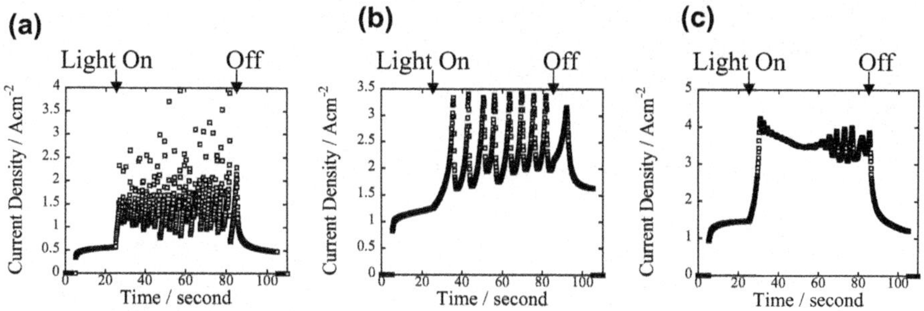

Fig. 9.6 Photocurrent oscillation observed for temperatures of **a** 77, **b** 193, and **c** 273 K by voltage-controlled measurements. Applied voltages are 63, 38, and 30 V, respectively

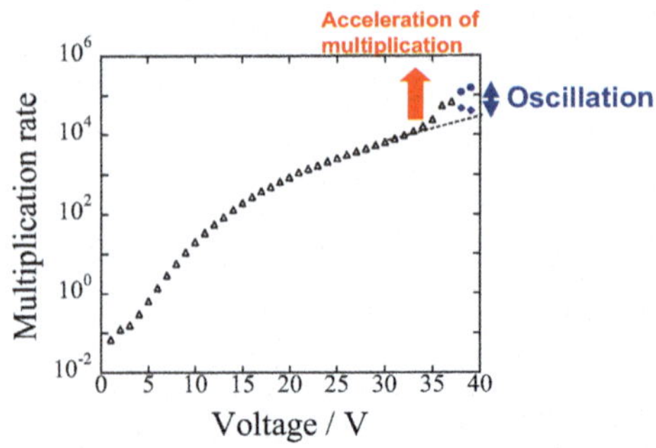

Fig. 9.7 Dependence of multiplication rate on the applied voltage at 5 K for the same cell of Figs. 9.4, 9.5, and 9.6

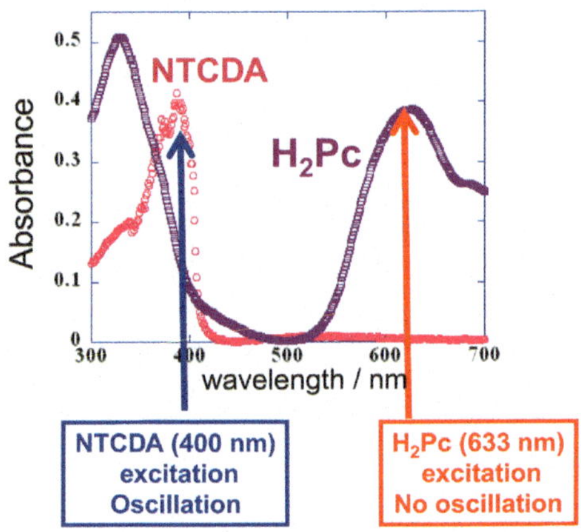

Fig. 9.8 Absorption spectra of NTCDA (48 nm) and H_2Pc (48 nm) films

little influence on the oscillation behavior. We believe that during the photocurrent oscillation, only a significantly small fraction of the applied voltage is distributed through the NTCDA film bulk, while most of the applied voltage is concentrated across the thin H_2Pc film.

9.7 Heterogeneous Accumulation Model

Photocurrent multiplication at the NTCDA/H_2Pc heterojunction occurs by electron injection from the valence band of H_2Pc to the conduction band of NTCDA (Fig. 9.1), which is induced by electric field concentration across the thin H_2Pc layer due to the accumulation of holes at the NTCDA/H_2Pc interface (Fig. 9.13). Holes are captured by the molecular blind alleys on the NTCDA crystal surface, as described in Chap. 5

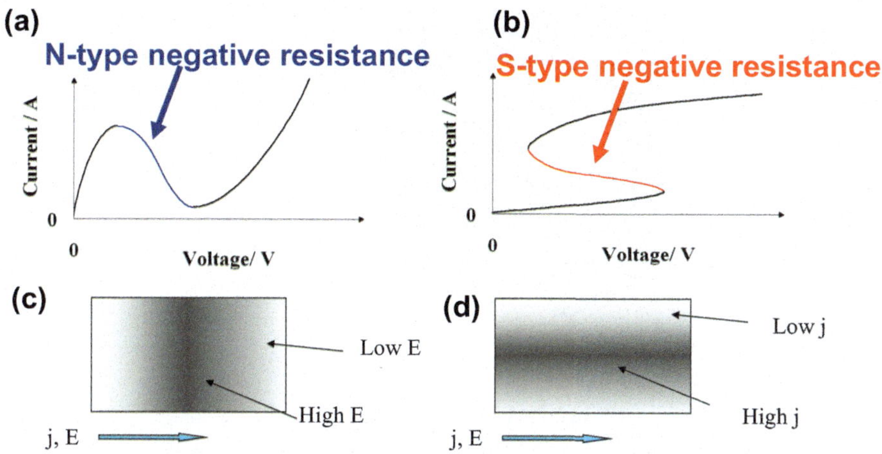

Fig. 9.9 **a** Current (I)-voltage (V) relationship of N-type negative resistance. **b** I-V relationship of S-type negative resistance. **c** Schematic of the electric field heterogeneity for N-type negative resistance. **d** Schematic of the current density heterogeneity for S-type negative resistance

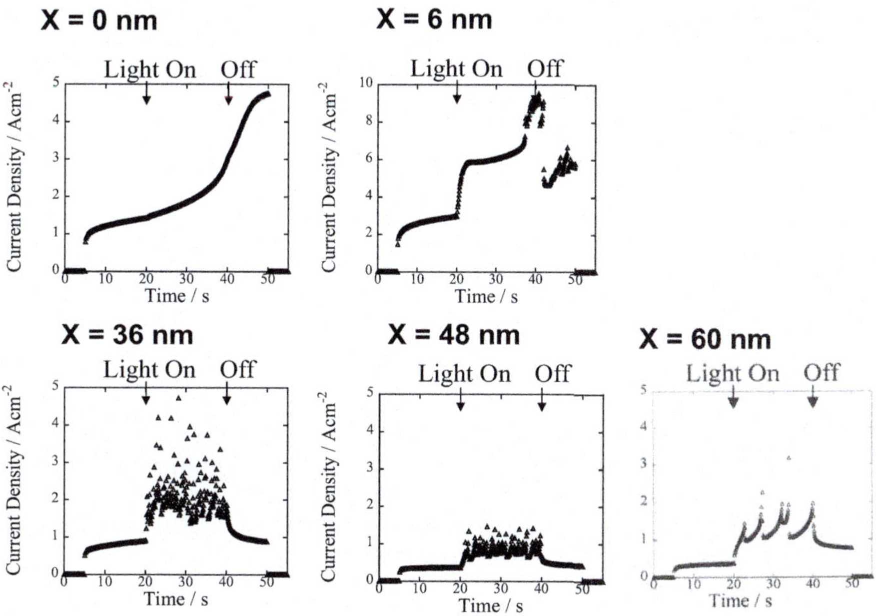

Fig. 9.10 Photocurrent response profiles for Fig. 9.2a cells having various H_2Pc thicknesses of 0, 6, 36, and 48 nm while keeping the same NTCDA thickness of 1 μm

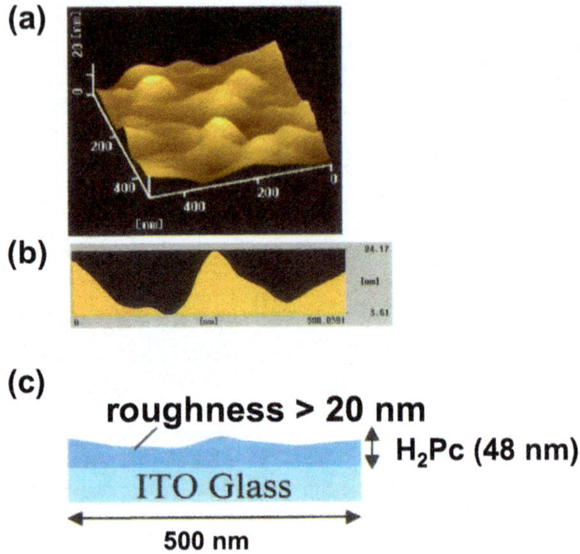

Fig. 9.11 **a** AFM image of H_2Pc film (48 nm) on ITO substrate. **b** Height profile of Fig. 9.11a. **c** Illustration of the cross section of H_2Pc film reproduced precisely based on Fig. 9.11b

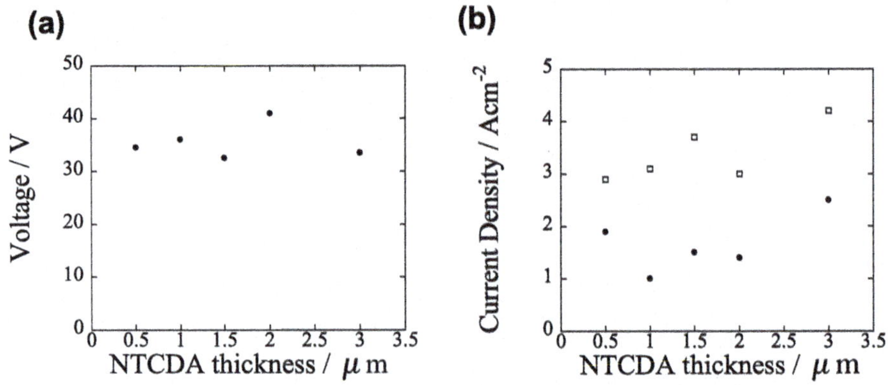

Fig. 9.12 **a** Dependence of the onset voltage of photocurrent oscillation on the NTCDA thickness. **b** Dependence of the oscillation region of photocurrent on the NTCDA thickness

(Fig. 9.13, lower right). In the low current density region on the S-shaped curve (Fig. 9.13, upper right), electron injection by the multiplication process occurs homogeneously because the low-concentration hole accumulation at the NTCDA/H_2Pc interface also occurs homogeneously (Fig. 9.13, left). Notably, the roughness of H_2Pc (in the order of several hundred nanometers) (Fig. 9.13, left) is significantly more than that of the molecular blind alleys (Fig. 9.13, lower right).

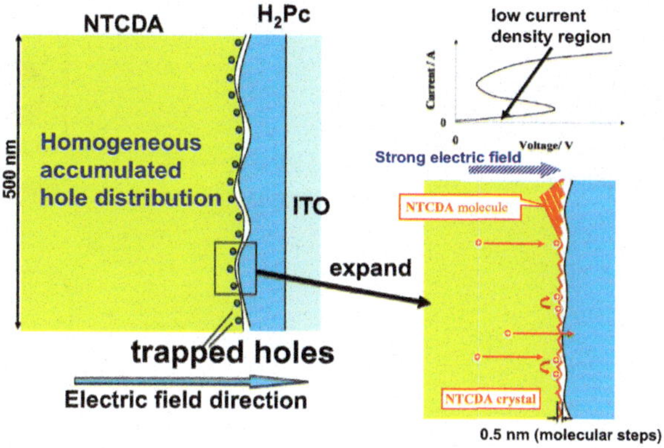

Fig. 9.13 Homogeneous hole accumulation at NTCDA/H_2Pc heterointerface in the low current density region. Molecular blind alley model is also shown (lower right)

We propose a heterogeneous accumulation model for current multiplication (Fig. 9.14). In the negative resistance region on the S-shaped curve (Fig. 9.14, right, red part), hole accumulation is accelerated in the thin H_2Pc region owing to the formation of a higher electric field perpendicular to the direction of the H_2Pc film thickness, relative to that in the thick H_2Pc region. Thus, electron injection is accelerated in the thin H_2Pc region. Consequently, spatially separated high current density regions (pink regions) and low-current–density regions (yellow regions) emerge (Fig. 9.14, left). The formation of the high current density region decreases the required voltage and leads to the formation of an S-type negative-resistance region.

In the voltage-controlled measurement, the spatial heterogeneity of accumulated holes results in the dynamic formation and disappearance of high and low current density regions in the S-type negative resistance region (see Sect. 9.5) (Fig. 9.14). Thus, current stability is lost and oscillation ensues oscillates. In this model, the spatial distribution heterogeneity of the accumulated holes is the origin of the S-shaped negative resistance and current oscillation.

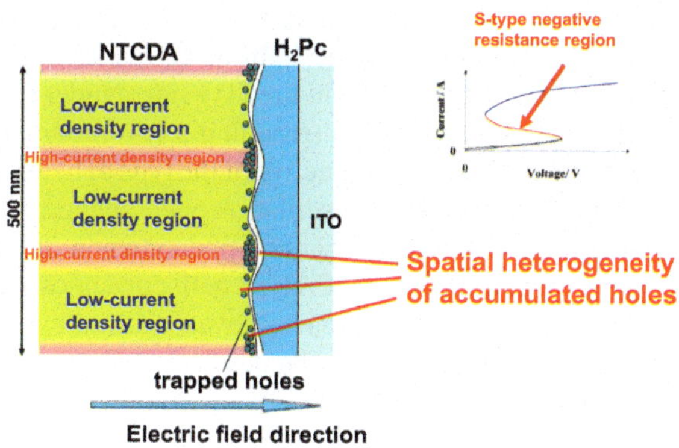

Fig. 9.14 Heterogeneous accumulation model for the S-type negative resistance region

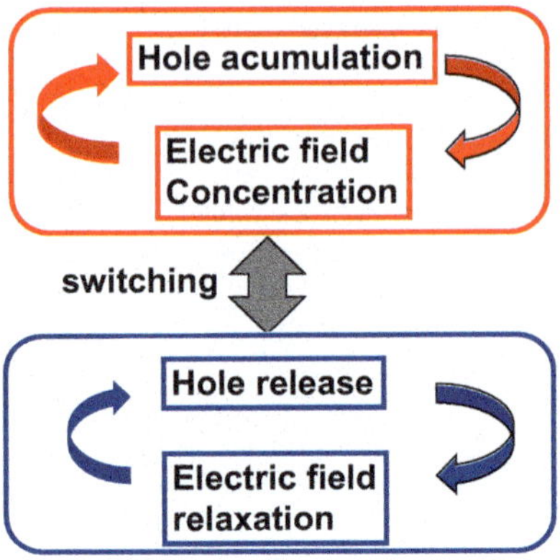

Fig. 9.15 Feedback effects between hole accumulation and electric field concentration and between hole release and electric field relaxation

9.8 Feedback Effect

The rise in the single waveform of the current oscillation accelerates at the inflection points; the switch from rise to decay occurs suddenly (Fig. 9.4c). In the experimental system, hole accumulation in the thin H_2Pc region caused further hole accumulation in the concentrated electric field, that is, the number of accumulated holes increased geometrically (Fig. 9.15). Conversely, hole release induces further hole release in a relaxed electric field, that is, the number of accumulated holes decreases geometrically (Fig. 9.15). These feedback mechanisms are inherently unstable and undergo switching between the two owing to the subtle heterogeneity of the spatial hole distribution. Such repeated switching results in dynamic current oscillations.

9.9 Generality

The NTCDA in the NTCDA/H_2Pc heterojunction was replaced by Me-PTC (Fig. 9.16a). Monochromatic light at 480 nm, which can be absorbed by the Me-PTC film, was irradiated from the ITO side. Under the voltage-controlled condition, the same current oscillation (reaching a current amplitude of 0.7 Acm^{-2}) could be observed (Fig. 9.16b). Irrespective of the type of heterojunction of Me-PTC/H_2Pc and NTCDA/H_2Pc, the current oscillation phenomenon could be observed. This indicates that current oscillation is a general phenomenon of organic heterojunctions.

In the case of the Me-PTC/H_2Pc cell, current oscillations were observed even for the dark current. This dark current was induced by the same multiplication process associated with the injection of holes from the positively biased Ag counter electrode. This result suggests that light is not strictly necessary to construct a practical organic DC-AC converter.

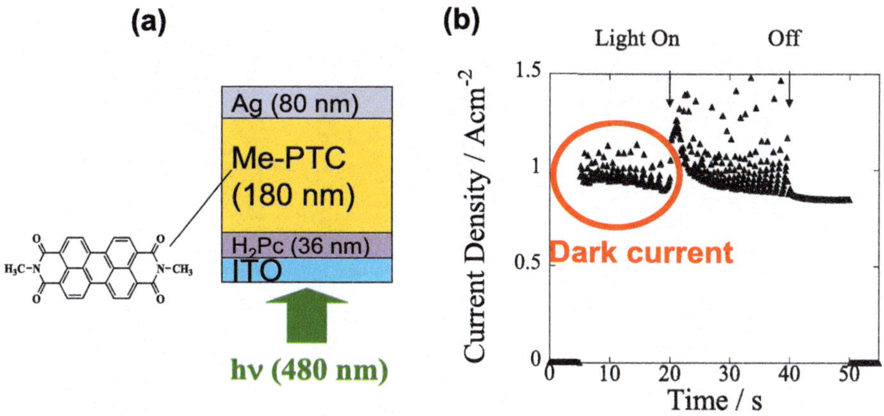

Fig. 9.16 **a** Structure of the cell having Me-PTC/H_2Pc heterojunction. Ag electrode was positively biased with respect to ITO electrode. **b** Current oscillation observed under the voltage-controlled condition

9.10 Removal of H_2Pc Roughness

9.10.1 Mechanical Pressing

The origin of the current oscillation was attributed to the thickness heterogeneity of the H_2Pc film. Therefore, the smoothening of the H_2Pc surface is expected to suppress current oscillation. To reduce the roughness, the H_2Pc film (60 nm) was directly pressed on a flat Si wafer at pressures of 40 and 80 MPa using a mechanical pressing machine for nanoimprinting (Fig. 9.17a) [14–17]. The color of the H_2Pc film in the pressed area changed as a result (Fig. 9.17b). Following H_2Pc pressing, NTCDA (1 μm) and Ag electrodes were deposited and the resulting photocurrent oscillation behavior was measured.

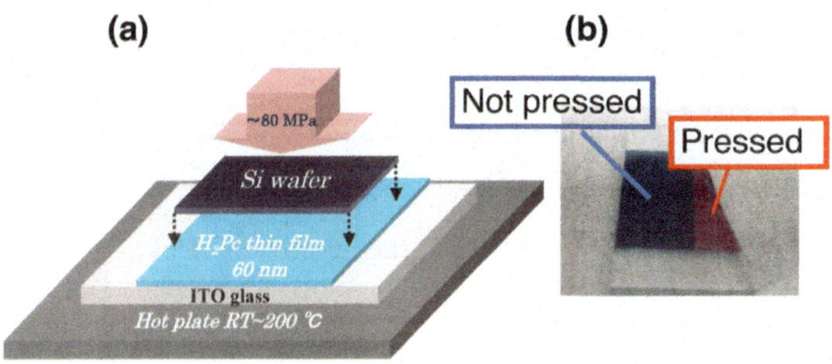

Fig. 9.17 **a** Mechanical pressing of H_2Pc film (60 nm) by the flat Si wafer using pressing machine. **b** Photograph of H_2Pc film having pressed and not pressed areas

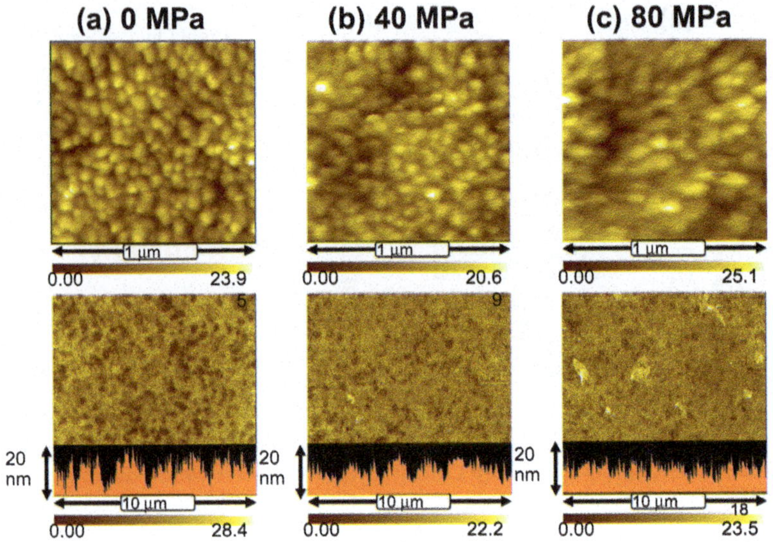

Fig. 9.18 AFM images of 1 μm × 1 μm and 10 μm × 10 μm area and the surface profiles of H_2Pc films (60 nm) pressed at the pressures of **a** 0, **b** 40, and **c** 80 MPa

9.10.2 H_2Pc Roughness

Figure 9.18a–c show the AFM images of the 1 μm × 1 μm and 10 μm × 10 μm areas, as well as the surface profiles of the H_2Pc following by pressing at pressures of 0, 40, and 80 MPa. As expected, an increase in the pressure corresponded to a reduction in the roughness of the H_2Pc film. Quantitatively, the average of surface roughness reduced from 2.7 to 2.0 and 1.7 nm at the pressures of 0, 40, and 80 MPa (Fig. 9.21, left vertical axis).

9.10.3 Oscillation Behavior and H_2Pc Roughness

Figure 9.19 shows the photocurrent oscillation observed by voltage-controlled measurements. The amplitude of the oscillation decreased, while the frequency of the oscillation increased with increasing applied pressure (Fig. 9.20). Moreover, the average roughness of the H_2Pc film and the amplitude of oscillation were simultaneously measured (Fig. 9.21). Figure 9.22 shows the dependence of the amplitude (black dots) and frequency (red dots) of oscillation on the average roughness. The amplitude increased with increasing average roughness. Conversely, the frequency decreased with increasing average roughness. Thus, the close relationship between the photocurrent oscillation and H_2Pc thickness heterogeneity (roughness) was confirmed.

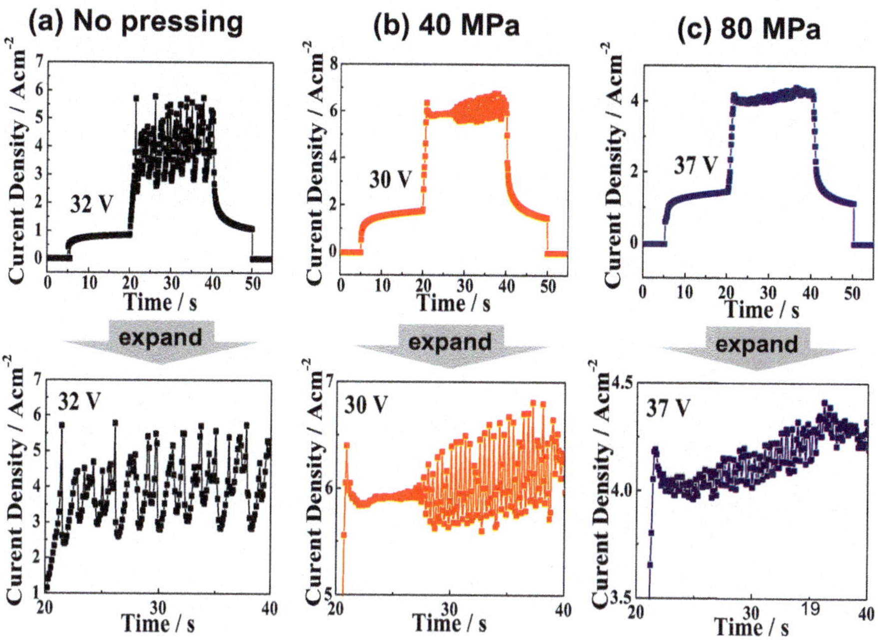

Fig. 9.19 Photocurrent oscillation observed by the voltage-controlled measurements. Enlarged figures focused around the oscillation region are also shown (lower). H_2Pc films (60 nm) were pressed at the pressures of **a** 0, **b** 40, and **c** 80 MPa

Fig. 9.20 Dependences of the amplitude (black squares) and frequency (red squares) on the applied pressure

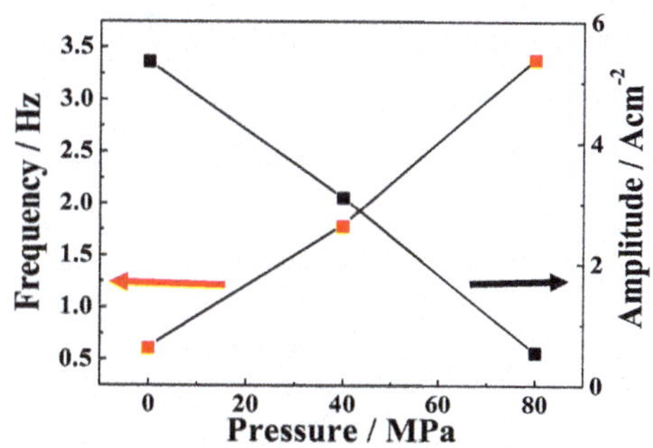

We believe that the increase in H_2Pc roughness, that is, the vertical direction of heterogeneity of the H_2Pc film (r) (from Figs. 9.23b to 9.23a) induces an increase in the heterogeneity of the spatial distribution of the number of trapped holes. Therefore, the heterogeneity of the multiplied photocurrent injected from H_2Pc to NTCDA (red arrows) increased, while the oscillation amplitude also increased (Fig. 9.22, black dots).

On the other hand, the decrease in the distance of the lateral direction of heterogeneity of the H_2Pc film (L) (from Fig. 9.23b, c) causes a more rapid development

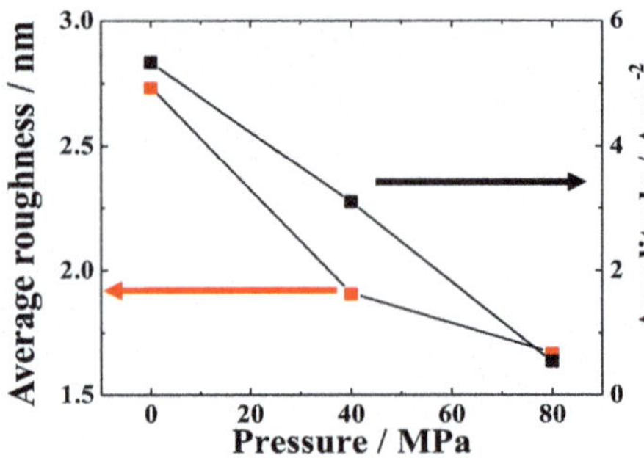

Fig. 9.21 Dependences of the average roughness of H_2Pc film (red squares) and the amplitude of oscillation (black squares) on the applied pressure

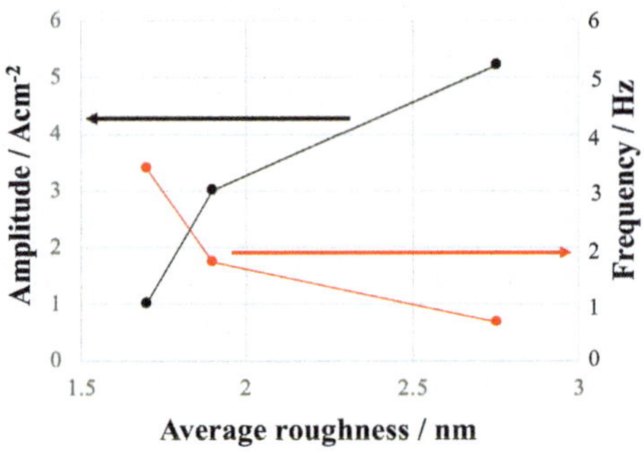

Fig. 9.22 Dependences of the amplitude (black dots) and frequency (red dots) of oscillation on the average roughness

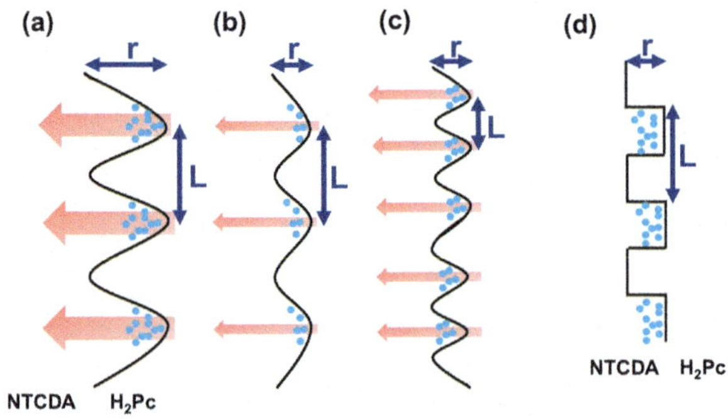

Fig. 9.23 Schematic of the roughness of NTCDA/H_2Pc heterointerface. Holes (light blue dots) are accumulated at traps on the NTCDA surface. Pink arrows indicate the injection of electrons via multiplication. **a** Large heterogeneity of H_2Pc thickness (roughness) (r). **b** Small heterogeneity of H_2Pc thickness (roughness) (r). **c** Smaller distance heterogeneity to the lateral direction of H_2Pc film (L). **d** Artificial nanostructure having the intentional r and L values. r and L have the dimensions of several nm and several hundred nm, respectively

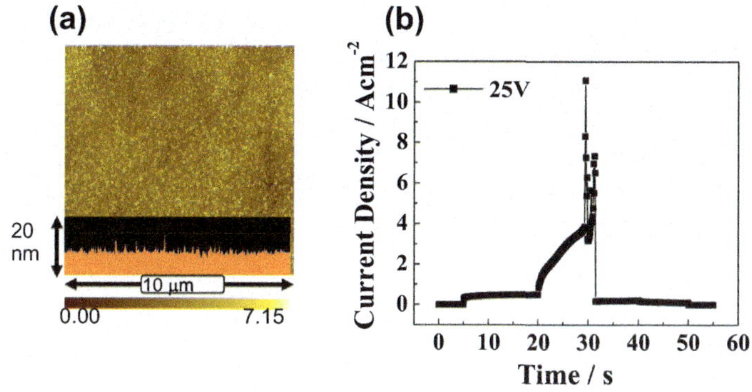

Fig. 9.24 **a** AFM image with height profile for H_2Pc film (60 nm) at the applied pressure of 40 MPa under 200 °C temperature. **b** Multiplied photocurrent response for the cell (Fig. 9.2a) incorporating H_2Pc film observed in Fig. 9.24a

of the heterogeneity in the spatial distribution of the number of trapped holes. Here, we suppose that a minute roughness is formed by mechanical pressing. Therefore, the frequency of the oscillation increased.

The above observations strongly suggest that the oscillation behaviors can be intentionally controlled by designing the r and L values of the nanoimprint mold with the shape shown in Fig. 9.23d. The removal of the roughness can be said to coincide with r = 0 nm in Fig. 9.23d. In this case, the disappearance of oscillation was already observed, as mentioned in Sect. 9.10.4.

9.10.4 Oscillation Disappearance

When pressing (40 MPa) was performed at an elevated temperature of 200 °C, the roughness of the H_2Pc film was almost completely eliminated (Fig. 9.24a). In this case, a slight oscillation was observed (Fig. 9.23b). This result provides clear evidence that the origin of current oscillation is the H_2Pc roughness, proving the validity of the proposed "Heterogeneous Accumulation Model" (Fig. 9.14).

9.10.5 Conclusion

The photocurrent oscillation phenomenon for multiplied photocurrent was observed for organic heterojunctions at temperatures between 5 and 288 K. The maximum amplitude of the oscillation reached a value of 3 Acm^{-2}. The oscillation frequency was 0.1–2 Hz. The photocurrent oscillation was induced by the S-shaped negative resistance. The heterogeneity of the H_2Pc film thickness, that is, the H_2Pc film roughness, was confirmed to be the origin of the S-shaped negative resistance. The formation of a heterogeneous distribution of accumulated holes induces high and low multiplied current density regions. Smoothening of the roughness by mechanical pressing causes the disappearance of current oscillation, which is clear evidence that the origin of the current oscillation is H_2Pc thickness heterogeneity. The current oscillation can be equated with a DC-AC conversion function for application to organic thyristor devices.

References

1. Hiramoto, M., Kumaoka, H., Yokoyama, M.: Photocurrent multiplication phenomenon at organic/organic heterojunction and application to optical computing device combining with organic electroluminescence. Synth. Met. **91**, 77–79 (1997)
2. Hiramoto, M., Nakayama, K., Sato, I., Kumaoka, H., Yokoyama, M.: Photocurrent multiplication phenomena at organic/metal and organic/organic interfaces. Thin Solid Films **331**, 71–75 (1998)
3. Hiramoto, M., Nakayama, K., Katsume, T., Yokoyama, M.: Field-activated structural traps at organic pigment/metal interfaces causing photocurrent multiplication phenomena. Appl. Phys. Lett. **73**, 2627–2629 (1998)
4. Hiramoto, M.: A concept of molecular blind alleys and organic multiplication-type photosensor. In: Fabrication of Organic Films and Application to Organic Devices, pp.70–79. CMC (2008) (in Japanese)
5. Wagner, H.J., Loutfy, R.O., Hsiao, C.: Purification and characterization of phthalocyanines. J. Mater. Sci. **17**, 2781–2791 (1982)
6. Esaki, L., Tsu, R.: IBM Research Note RC-2418 (1969)
7. Esaki, L., Tsu, R.: IBM Res Develop. **14**, 61 (1970)
8. Esaki, L.: A perspective in superlattice development. In: Physics and Application of Semiconductor Superlattices. Physical Society of Japan, Baifu-kan, pp. 253–300 (1984) and references therein
9. Esaki, L.: The evolution of nanoscale quantum effects in semiconductor physics. Nanostructured Mater. **12**, 1–8 (1999)
10. Tsu, R.: Applying the insight into superlattices and quantum wells for nanostructures: low-dimensional structures and devices. Microelectronics J. **38**, 959–1022 (2007)
11. Sawano, F., Terasaki, I., Mori, H., Mori, T., Watanabe, M., Ikeda, N., Nogami, Y., Noda, Y.: An organic thyristor. Nature **437**, 522–524 (2005)
12. Shoel, E.: Nonequilibrium Phase Transitions in Semiconductors. Springer, Berlin, Heidelberg (1987)
13. Takahashi, H.: Optically controllable S-type negative resistance presented by a combinational connection of photocoupled FET's. IEEE Trans. Electron Dev. **30**, 647–652 (1983)
14. Schift, H.: Nanoimprint lithography: 2D or not 2D?A review. Appl. Phys. A **121**, 415–435 (2015)

15. Chou, S.Y., Krauss, P.R., Renstrom, P.J.: Imprint of sub25 nm vias and trenches in polymers. Appl. Phys. Lett. **67**, 3114–3116 (1995)
16. Hirai, Y.: Polymer science in nanoimprint lithography. J. Photopolym. Sci. Tech. **18**, 551–558 (2005)
17. Sreenivasan, S.V.: Nanoimprint lithography steppers for volume fabrication of leading-edge semiconductor integrated circuits. Microsyst. Nanoeng. **3**, 17075 (19 pages) (2017)

Chapter 10
Avalanche Multiplication in Perylene Molecular Crystals

10.1 Background

In the last decade, band conductive organic single crystals with high hole mobilities of up to 43 cm^2V^{-1} s^{-1}, measured in a field-effect transistor configuration, have been reported [1–5]. Due to strong π–π interactions, a high mobility is observed in the direction parallel to the crystal surface (Fig. 10.1b). In his early work in 1985 [6, 7], Karl reported extremely high carrier mobilities of up to 100 cm^2V^{-1} s^{-1} for highly purified molecular crystals, such as naphthalene and perylene, probed using time-of-flight (TOF) measurements when the crystals were cooled below 30 K (Fig. 10.2). These mobility values are close to those of silicon crystals [8] and much larger than those of amorphous selenium, which is the only amorphous material that shows the avalanche multiplication effect [9–11].

10.2 Motivation

The perylene molecular crystals show a high electron mobility of 100 cm^2V^{-1} s^{-1} at low temperatures, i.e., below 30 K [6]. By using an electric field of 1.5×10^6 Vcm^{-1} and a mobility (μ) of 100 cm^2V^{-1} s^{-1}, the kinetic energy [$(1/2)mv^2$] of a carrier is calculated to be 6.3 eV, which is significantly higher than the bandgap energy of a perylene crystal (2.5 eV). This observation implies that impact ionization occurs in high-mobility molecular crystals, such as perylene (Figs. 10.1 and 10.2). Therefore, we studied avalanche multiplication in perylene single crystals at temperatures below 30 K.

M. Hiramoto, *Organic Photocurrent Multiplication*, Electronic Materials: Science & Technology, https://doi.org/10.1007/978-981-99-1237-7_10

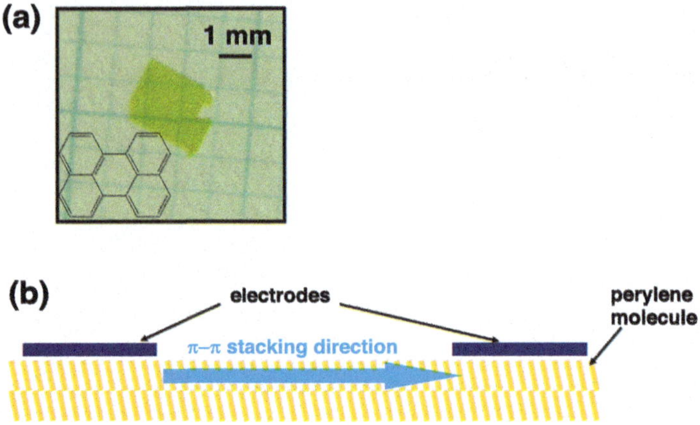

Fig. 10.1 **a** Image of a perylene single crystal. **b** Schematic of π–π stacking parallel to the crystal surface, which shows the high carrier mobility

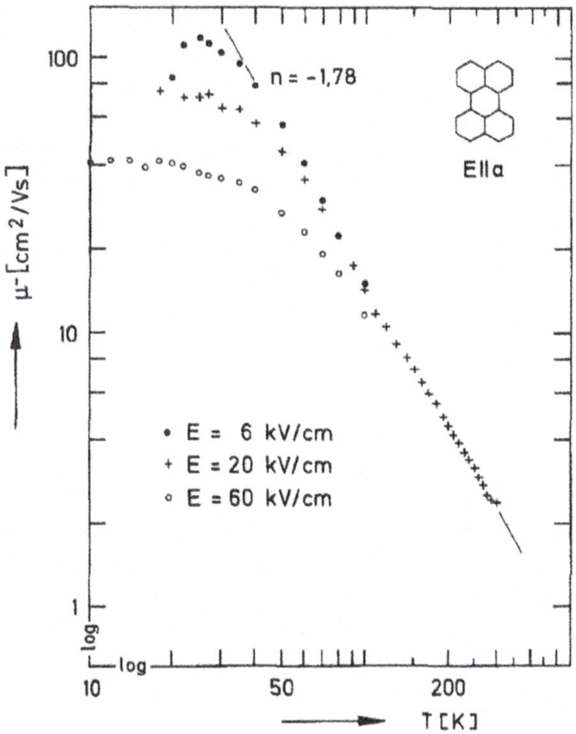

Fig. 10.2 Electron mobilities in perylene single crystal for the electric field E parallel to the crystallographic surface direction for different field strengths (marked by different symbols) between 6 and 60 $kVcm^{-1}$. Reproduced with the permission of [6]. © 1985 Springer-Verlag

10.3 Photocurrent Multiplication Rate in Perylene Single Crystals

At an applied electric field of 1.4×10^6 Vcm^{-1} and low temperatures of 5 K, photocurrent multiplication increases 120-folds. The multiplication rate increases with an increase in the carrier-traveling distance [12].

10.3.1 Perylene Single Crystals

A sample of perylene (Tokyo Kasei, Co.) was purified by vacuum-train sublimation twice at 2×10^{-2} Torr pressure, prior to crystal growth [13]. Single crystal growth was observed under flowing N_2 gas at 1 atm pressure using a temperature-gradient furnace [14]. The purified samples were injected at 250 °C, and the crystals were grown at approximately 130 °C. The grown perylene crystals were typically 5 mm × 5 mm in size and 30 μm in thickness (Fig. 10.1a). X-ray diffraction measurements confirmed that the obtained crystals have the same crystal structure as that of α-perylene (Fig. 10.10) [6], in which the π–π stacking is parallel to the crystal surface.

10.3.2 Surface-Type Cells

As perylene crystals show high carrier mobility along the π-stacking direction parallel to the surface (Figs. 10.1 and 10.2) [6, 15], surface-type cells were fabricated (Figs. 10.3). Pairs of Au electrodes were vacuum deposited using metal masks with tungsten wires having diameters corresponding to the required electrode spacings (L) of 8, 5, and 2.5 μm, as shown in Fig. 10.3c–e, respectively.

The photocurrent was measured by applying a voltage to the electrodes under 10^{-5} Pa at 5 K in a helium cryostat (Daikin Co. Ltd., PS24SS) that was evacuated by a turbo molecular pump (Alcatel, ATP-80) under 400-nm light irradiation from a GaN diode laser. The ratio of the number of carriers flowing through the cell, which is calculated using the photocurrent obtained by subtracting the dark current from the total current under light irradiation, to the number of photons absorbed by the crystal between electrodes (Fig. 10.3, brown area) was used to calculate the photocurrent quantum efficiency, i.e., the multiplication rate. The current density was calculated by assuming that the photocurrent flowed within 1 μm of the crystal surface.

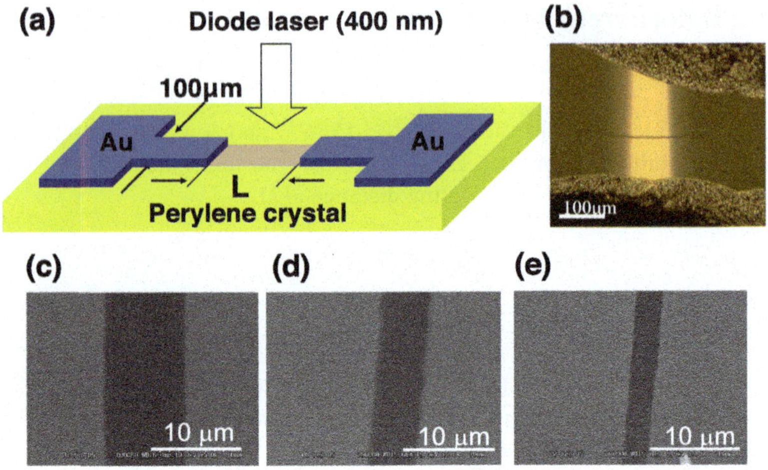

Fig. 10.3 **a** Surface-type cell. L denotes the distance between the electrodes, i.e., the carrier-traveling distance. **b** Image of a cell taken by optical microscope. The bright yellow parts are Au electrodes separating a gap at the center. Electrodes are connected to Ag paste seen at up and down. SEM images of cells with **c** L = 8 μm, **d** L = 5 μm, and **e** L = 2.5 μm

10.3.3 Dark Current

Figure 10.4a shows the dependence of the current density on the applied voltage in the dark. The dark current is effectively suppressed by lowering the temperature from 300 to 5 K. Based on the energy diagram (Fig. 10.4b) [15], hole injection from the Au electrode is considered the dominant source of dark current because of the lower hole injection barrier value (0.3 eV) than the considerable electron injection barrier value (2.2 eV). We conclude that the hole injection is effectively suppressed at 5 K. Therefore, we investigated the multiplication characteristics of crystals at 5 K.

10.3.4 Multiplication Characteristics

Figure 10.5a shows the multiplication-rate dependence of a cell with L = 5 μm on the applied voltage and electric field at 5 K. The multiplication rate exceeded unity at the applied voltage of 500 V, i.e., at the applied electric field (E) of 1.0×10^6 $\mathrm{Vcm^{-1}}$ (Fig. 10.5b). The multiplication rate steeply increased by 120-folds at 1.5×10^6 $\mathrm{Vcm^{-1}}$.

Figure 10.6a shows the enlarged plot of multiplication-rate dependence on the applied electric field at 5 K for surface-type cells with electrode spacings of 2.5, 5, and 8 μm (curves A, B, and C, respectively). At approximately 1.0×10^6 $\mathrm{Vcm^{-1}}$, the multiplication rate exceeded unity for all cells and attained approximately 10-folds of its value at 1.2×10^6 $\mathrm{Vcm^{-1}}$. The spectral dependence of the multiplied photocurrent

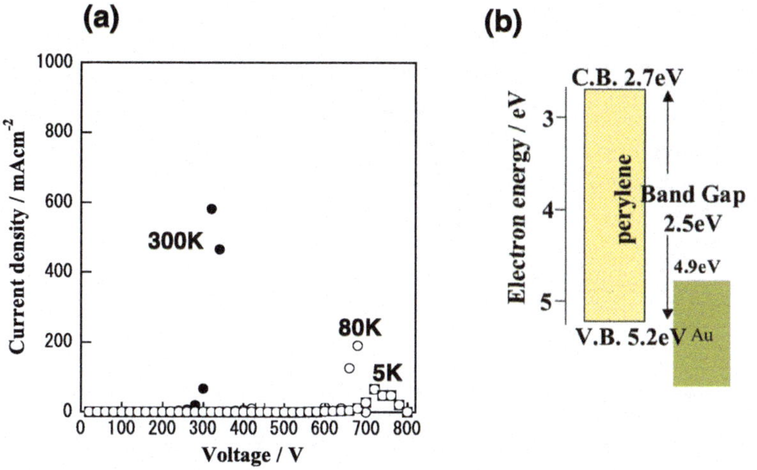

Fig. 10.4 **a** Dependence of the current density of dark current on the applied voltage at 300, 80, 5 K for the surface-type cell with L = 5 μm. **b** Energy diagram of perylene and Au electrode. CB and VB denote the conduction band and valence band, respectively

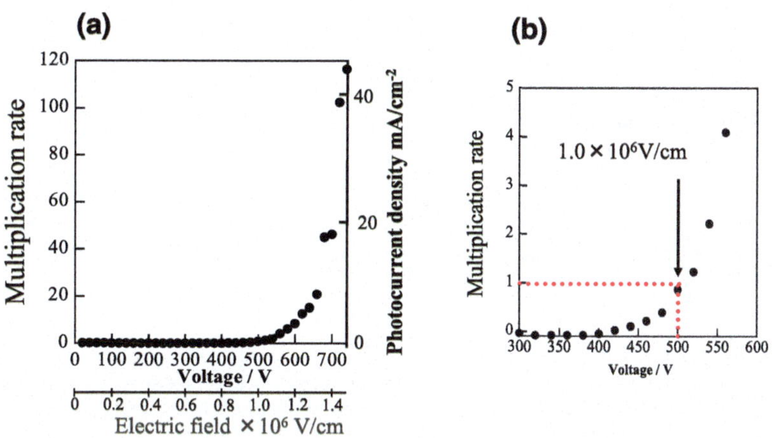

Fig. 10.5 **a** Dependence of the multiplication rate (left vertical axis) and photocurrent density (right vertical axis) on the applied voltage and electric field (E) for L = 5 μm cell. A 400-nm diode laser (0.95 mWcm^{-2}) irradiated the cell at 5 K. **b** Magnified image of Fig. 10.4a of the area around the multiplication rate of 1 (red dotted line)

was in good agreement with the shape of the absorption spectrum of the perylene crystal (Fig. 10.7a). The multiplied photocurrent increases proportionally with light intensity (Fig. 10.7b). The photoresponse profile of the multiplied photocurrent was reproduced multiple times and was unaffected by the preceding measurements. These features make the device suitable for its application as a photodetector. Thus, we

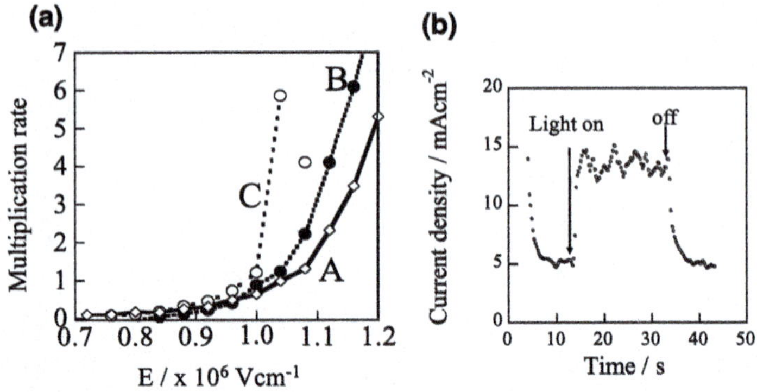

Fig. 10.6 **a** Dependence of the multiplication rate on the applied electric field (E) for the cells with L = 2.5 (curve A), 5 (curve B), and 8 μm (curve C). **b** Photoresponse profile of the multiplied photocurrent at $E = 1.32 \times 10^6$ Vcm^{-1}. The multiplication rate is approximately 20-folds. A 400-nm diode laser (0.95 $mWcm^{-2}$) irradiated the cells at 5 K

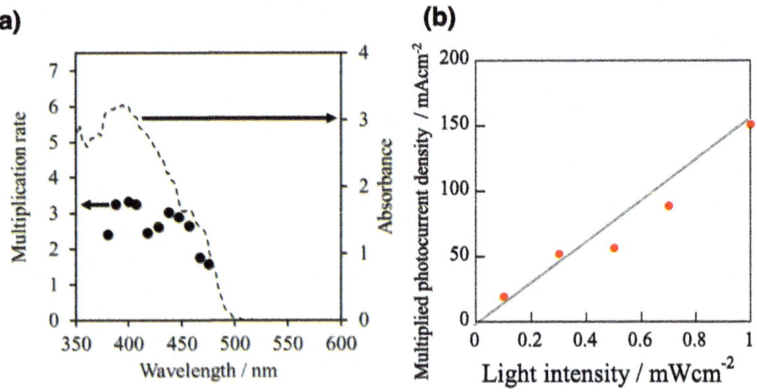

Fig. 10.7 **a** Spectral dependence of the multiplication rate (black dots). The absorption spectrum of a perylene single crystal is also indicated (broken curve). **b** Light intensity dependence of the multiplied photocurrent at $E = 1.04 \times 10^6$ Vcm^{-1}. A 400-nm diode laser irradiated the cells at 5 K

observed the multiplication of photocurrent in perylene single crystals using the surface-type cells.

10.3.5 Photoresponse

The multiplied photocurrent is always accompanied by random oscillations (Fig. 10.6b). This phenomenon is a typical feature of the avalanche effect. The present phenomenon is only observed at a very low temperature of 5 K, at which the electron

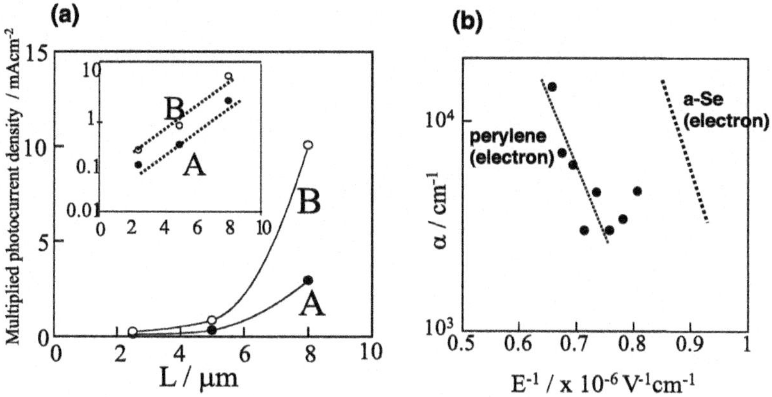

Fig. 10.8 **a** Dependence of multiplied photocurrent density (I_{ph}) on carrier-traveling distance (L) at a constant electric field (E) of 1.0×10^6 Vcm^{-1} (curve A) and 1.08×10^6 Vcm^{-1} (curve B). Inset: Semi-logarithmic plot. **b** Ionization rate (α) versus the inverse value of applied electric field (E^{-1})

mobility in the perylene crystal reached its maximum (Fig. 10.2) [6]. These features suggest that the observed multiplication is due to impact ionization.

10.3.6 Carrier-Traveling Distance

The most important feature of the multiplication observed for the perylene single crystals is that the multiplication rate for the same electric field increases with increasing electrode spacing from 2.5 to 8 µm, which corresponds to the carrier-traveling distance (L) (Fig. 10.3a). Figure 10.8a shows the dependence of the multiplied photocurrent density on L at a constant electric field of 1.0×10^6 and 1.08×10^6 Vcm^{-1} (curves A and B, respectively). Evidently, the multiplied photocurrent density rapidly increases with increasing L [12]. This increase is a unique feature of impact ionization (Fig. 10.9) [9].

10.4 Impact Ionization

Photogenerated electrons in perylene single crystals are accelerated along the crystal surface under a strong electric field exceeding 1.0×10^6 Vcm^{-1} (Fig. 10.9). When an electron has more energy than the threshold energy for impact ionization, an electron–hole pair is generated by the collision of an accelerated electron with a crystal lattice. Through repeated impact ionization, the number of electrons increases in geometric progression, i.e., from 1 to 2, 2 to 4, and 4 to 8, by the

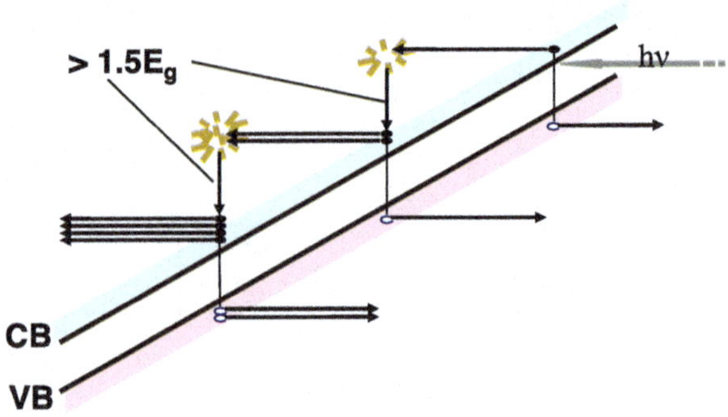

Fig. 10.9 Schematic of the mechanism of avalanche multiplication. The open and closed dots represent the holes and electrons, respectively. VB, CB, and Eg denote the valence band, conduction band, and bandgap, respectively

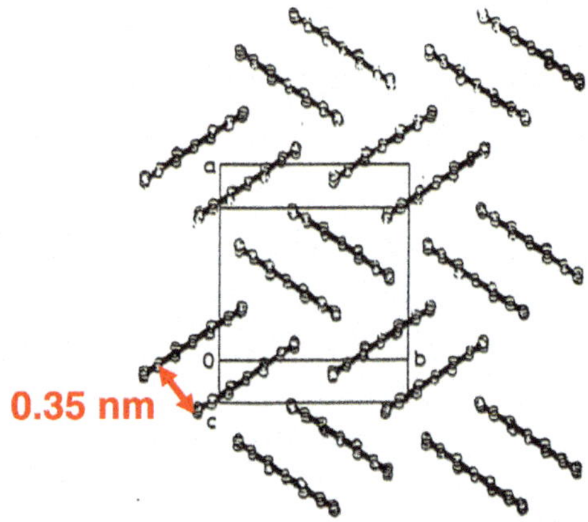

Fig. 10.10 Crystal structure of α-perylene. Reproduced with the permission of [6]. © 1985 Springer-Verlag

first, second, and third ionizations, respectively (Fig. 10.9). This process results in the multiplication of the photocurrent. This mechanism means that the longer carrier-traveling distance corresponds to more collisions (impact ionizations). Thus, the multiplied photocurrent rapidly increases with L. Only the avalanche mechanism enhances the traveling distance [9]. All other multiplication processes under high electric fields, such as the photoinduced tunneling injection of electrons at the organic/metal interface, discussed in other chapters of this book [16, 17], cannot explain the enhanced traveling distance. Moreover, the observed high-speed photoresponse confirms that the present multiplication is not caused by the photoinduced tunneling at the organic/metal interface, which exhibits a very slow photoresponse lasting several tens of seconds [16, 17]. Thus, we conclude that the present multiplication of the photocurrent is caused by avalanche multiplication through impact ionization.

The threshold energy of impact ionization is 1.5 Eg (where, Eg represents bandgap energy), which is based on the conservation laws of energy and carrier momentum [18]. Using the Eg of a perylene single crystal (2.5 eV), the threshold energy (1.5 Eg) of 3.75 eV is obtained. To initiate impact ionization, the carrier should be accelerated to attain a kinetic energy (E_{kin}) larger than 3.75 eV. The kinetic energy of a carrier (E_{kin}) is expressed as:

$$E_{kin} = (1/2)mv^2 = (1/2)m(\mu E)^2 \quad (10.1)$$

Here, m, v, μ, and E denote the carrier mass, carrier velocity, carrier mobility, and electric field, respectively. The electron mobility at 5 K is 100 cm^2V^{-1} s^{-1} [6]. At $E = 1.5 \times 10^6$ Vcm^{-1}, the calculated E_{kin} for an electron is 6.3 eV, which exceeds the threshold energy ($1.5E_g$) of 3.75 eV. Therefore, we believe that only electrons can cause impact ionization under the present experimental conditions, as shown in Fig. 10.9. The calculated E_{kin} of 3.76 eV coincides with the threshold energy ($1.5E_g$) of 3.75 eV at $E = 1.15 \times 10^6$ Vcm^{-1}, which is slightly larger than the threshold electric field at which the multiplication rate exceeds unity (1.0×10^6 Vcm^{-1}) (Fig. 10.5b). This coincidence further supports the impact ionization mechanism. Notably, the free electron mass, which is used in the calculations above, is not significantly different from the effective mass, which is reported as 0.9 in perylene [17].

10.5 Ionization Rate

In the case of avalanche multiplication, the multiplied photocurrent is proportional to exp(αL) [9]. Here, α is the ionization rate, i.e., the number of impact ionizations per unit distance. The observed dependence of the multiplied photocurrent on L (Fig. 10.8a) increases rapidly. Moreover, they exhibit a linear relationship in a semi-logarithmic plot (Fig. 10.8a, inset). At applied electric fields of 1.0×10^6 and 1.08×10^6 Vcm^{-1} (Fig. 10.8a, curves A and B, respectively), α values of 5.9×10^3 and 8.0×10^6 Vcm^{-1} were obtained from the slopes. Figure 10.8b shows the dependence of the ionization rate (α) on the inverse value of the applied electric field (E^{-1}), attaining 1.5×10^6 Vcm^{-1}. Values ranging from 10^3 to 10^4 cm^{-1} were obtained. These values are comparable to a values reported for amorphous selenium (Fig. 10.8b, broken line) [9].

10.6 Mean Free Path

In the case of avalanche multiplication in inorganic semiconductors, the straight line in Fig. 10.8b is defined as $\exp(-\varepsilon_i/eEl_0)$ [9]. Here, ε_i is the threshold energy for carrier ionization ($1.5E_g$), e is the elementary charge, E is the field strength, and l_0 is the mean free path of the carriers. For electrons in perylene, l_0 of 1.5 nm was obtained, which was comparable to the l_0 value of 3.2 nm reported for perylene single crystals based on the band model [19]. The mean free path (l_0) of 1.5 nm, which is 4.3 times the spacing of 0.35 nm in $\pi - \pi$ stacking in the perylene single crystal (Fig. 10.9), suggests that an electron can move through 4 perylene molecules without collision. This suggests the band conductive character of the perylene single crystal.

The occurrence of the impact ionization implies the existence of free carriers in the perylene single crystals because effective acceleration is only possible for free carriers in band conductive semiconductors. This phenomenon suggests that the perylene single crystals are band conductive semiconductors in the low-temperature region. Moreover, this conclusion is consistent with the fact that the electron mobility of perylene single crystals increases with decreasing temperature, which is a typical behavior of band conductive semiconductors [20, 21].

10.7 Conclusion

Avalanche multiplication of the photocurrent in perylene molecular crystals was discovered [12]. Other organic crystals and films with band conductive nature, such as rubrene [1], naphthalene [6, 7], and C8-BTBT [3, 4], might also exhibit avalanche multiplication. Notably, the type of multiplication phenomenon reported in this chapter is different from those based on the interfacial tunneling injection, which are described in the other chapters of this book. We believe that this work can lead to the development of avalanche photodetectors made of organic molecular crystals and films that show band conductive nature at room temperatures.

References

1. Takeya, J., Yamagishi, M., Tominari, Y., Hirahara, R., Nakazawa, Y.: Very high-mobility organic single-crystal transistors with in-crystal induction channels. Appl. Phys. Lett. **90**, 102120 (3 pages) (2007)
2. Minemawari, H., Yamada, T., Matsui, H., Tsutsumi, J., Haas, S., Chiba, R., Kumai, R., Hasegawa, T.: Inkjet printing of single-crystal films. Nature **475**, 364–367 (2011)
3. Yuan, Y., Giri, G., Ayzner, A.L., Zoombelt, A.P., Mannsfeld, S.C.B., Chen, J., Nordlund, D., Toney, M.F., Huang, J., Bao, Z.: Ultrahigh mobility transparent organic thin film transistors grown by an off-centre spin-coating method. Nat. Commun. **5**, 3005 (5 pages) (2014)
4. Haas, S., Takahashi, Y., Takimiya, K., Hasegawa, T.: High performance dinaphtho-thieno-thiophene single crystal field-effect transistors. Appl. Phys. Lett. **95**, 022111 (2009)
5. Chesterfield, R.J., McKeen, J.C., Newman, C.R., Ewbank, P.C., da Silva Filho, D.A., Brédas, J.-L., Miller, L.L., Mann, K.R., Frisbie, C.D.: Organic thin film transistors based on N-alkyl perylene diimides: charge transport kinetics as a function of gate voltage and temperature. J. Phys. Chem. B **108**, 19281–19292 (2004)
6. Warta, W., Stehle, R., Karl, N.: Ultrapure, high mobility organic photoconductors. Appl. Phys. A **36**, 163–170 (1985)
7. Warta, W., Karl, N.: Hot hole in naphthalene: high, electric-field-dependent mobilities. Phys. Rev. B **32**, 1172–1181 (1985)
8. Sze, S.M.: Physics of Semiconductor Devices, pp. 675–683. Wiley-Interscience, New York (1969)

9. Takasaki, Y., Tsuji, K., Hirai, T., Maruyama, E., Tanioka, K., Yamazaki, J., Shidara, K., Taketoshi, K.: Avalanche multiplication of photo-generated carriers in amorphous semiconductor and its application to imaging device. Mater. Res. Soc. Symp. Proc. **118**, 387–397 (1988)
10. Taketoshi, K., Tanioka, K., Andoh, F., Yamazakii, J., Takasaki, Y.: Analyses of noise in a highly sensitive image device. Jpn. J. Appl. Phys. **28**, 178–186 (1989)
11. Tsuji, K., Ohshima, T., Hirai, T., Gotoh, N., Taniokata, K., Shidara, K.: Ultra-high-sensitive image pickup tubes using avalanche multiplication in a-Se. Mater. Res. Soc. Symp. Proc. **219**, 507–518 (1991)
12. Hiramoto, M.: Avalanche multiplication-type photosensors using ultra-thin molecular crystals and their fabrication method. Jpn. Patent, No. 3976700 (2007) (in Japanese)
13. Wagner, H.J., Loutfy, R.O., Hsiao, C.: Purification and characterization of phthalocyanines. J. Mater. Sci. **17**, 2781 (1990)
14. Laudise, R.A., Kloc, Ch., Simpkins, P.G., Siegrist, T.: Physical vapor growth of organic semiconductors. J. Crystal Growth **187**, 449–454 (1998)
15. Hiramoto, M., Tomioka, A., Suemori, K., Yokoyama, M.: Formation of ohmic contacts to perylene molecular crystals. Appl. Phys. Lett. **85**, 1852–1854 (2004)
16. Hiramoto, M., Nakayama, K., Katsume, T., Yokoyama, M.: Molecular-sized structural trap at organic-metal interface and photocurrent multiplication phenomenon. In: Salaneck, W.R., Seki, K., Kahn, A., Pireaux, J.-J. (eds.) Chapter 18. Conjugated Polymer and Molecular Interfaces, pp. 585–612. Marcel Dekker Inc., New York (2001)
17. Hiramoto, M., Imahigashi, T., Yokoyama, M.: Photocurrent multiplication in organic pigment films. Appl. Phys. Lett. 64, 187–189 (1994)
18. Hamaguchi, K., Tanida, K.: Physics of Semiconductor Devices, p. 141 (1990)
19. Karl, N., Marktanner, J., Stehle, R., Warta, W.: High-field saturation of charge carrier drift velocities in ultrapurified organic semiconductors. Synth. Metals **41–43**, 2473–2481 (1991)
20. Kikuchi, M., Izawa, S., Rai, N., Hiramoto, M.: Very low activation energy for carrier generation of surface doped organic single crystals observed by Hall effects. Appl. Phys. Lett. **115**, 113301 (4 pages) (2019)
21. Minder, N.A., Ono, S., Chen, Z., Facchetti, A., Morpurgo, A.: Band-like electron transport in organic transistors and implication of the molecular structure for performance optimization. Adv. Mater. **24**, 503–508 (2012)

Chapter 11
Progress in Organic Photocurrent Multiplication

11.1 Aim of This Chapter

Following the authors' work in 1994, described in this book, research on organic photocurrent multiplication (OPM) mushroomed after 2000. Summarizing the progress in multiplication research thus far will help predict the future progress of OPM. In this chapter, a brief history of OPM is outlined and representative studies closely related to the fundamental mechanism are described.

11.2 Historical Outline

Table 11.1 summarizes representative studies on organic photocurrent multiplication (OPM) devices from 1994 to 2017 that appeared in reviews [1, 2]. Studies of Au/Me-PTC/Au [3] and Au/NTCDA/Au [4] highlighted by red rectangles are the authors' discovery of multiplication phenomena in the first reported OPM systems comprising vacuum-deposited films and organic single crystals.

The studies highlighted by pale yellow in Table 11.1 [1] focus on OPM devices employing organic films of small molecules deposited by vacuum evaporation. The reproducibility of photocurrent multiplication using small-molecule films was confirmed by Huang and Yang [5]. They reported that the ITO/PEDOT:PSS/C_{60}/BCP/Al cell exhibited a multiplication rate of 5,000%, that is, a multiplication rate of 50-fold. Here, PEDOT:PSS is 3,4-ethylenedioxythiophene)poly(styrenesulfonate), which acts as the conducting electrode. They showed that both the disordered structure of C_{60} and the charge-trapping effect at the C_{60}/PEDOT:PSS interface contribute to the photomultiplication effect.

The studies in Table 11.1 [1] that are not highlighted by color focus on OPM devices employing organic polymer films deposited by spin-coating from organic solvents. The progress in the development of OPM devices has been strongly influenced by that in the development of organic solar cells. Photocurrent multiplication

M. Hiramoto, *Organic Photocurrent Multiplication*, Electronic Materials: Science & Technology, https://doi.org/10.1007/978-981-99-1237-7_11

Table 11.1 Representative studies on organic photocurrent multiplication devices from 1994 to 2017. Yellow shading and no shading correspond to the studies using small-molecules and polymers, respectively. Blue reference numbers correspond to the references in this chapter. Reference numbers in the Table correspond to the references appearing in the review [1]. Reproduced with the permission of [1]. © 2019, WILEY–VCH Verlag GmbH

Year	Materials [device structure]	Location of traps	Response region [nm]	EQE [%]	Response speed [s]	D^* [Jones]	Features	Ref	
1994	Au/Me-PTC/Au	Structural traps (formed by the imperfect organic/metal contact)	350–700	10 000	–	–	The first reported PM type OPDs	36	[3]
1999	Au/arylamino-PPV/Al		450–520	2000	–	–	PM type polymer photodetectors	37	[6]
2002	Au/NTCDA/Au		300–450	20 000	0.5	–	PM phenomenon in organic single crystal	54	[4]
2005	MEH-PPV:PbSe (Vertical)	In active layer	400–600	150	–	–	PbSe QDs ($PC_{61}BM$) as traps in polymer matrix	48	
2007	MEH-PPV:$PC_{61}BM$ (or PbSe) (Vertical)		400–600	10 000	$\approx 10^{-3}$	–		49	
2008	P3HT:$PC_{61}BM$:CdTe (Vertical)		350–700	8000	$\approx 10^{-5}$	–	CdTe NPs as traps in ternary blend film	39	
2012	P3HT:ZnO or PVK:ZnO (Vertical)		300–650 300–380	$\approx 10^{5}$	$\approx 10^{-4}$	2.5×10^{14} (−9 V) 3.4×10^{15}	ZnO NPs as traps	40	
2010	P3HT:$PC_{61}BM$:Ir-125 (Vertical)		300–1050	7000	–	–	NIR Dye (Ir-125 & Q-switch 1) as traps in ternary blend film; meanwhile extend spectral response to NIR region.	41	
2012	P3HT:$PC_{61}BM$:Ir-125:Q-switch 1 (Vertical)		300–1200	5500	–	–		60	
2014	PbS:P3HT:$PC_{61}BM$:ZnO (Vertical)		320–1000	1624	$\approx 10^{-4}$	8.3×10^{11} (−4 V)	ZnO QDs as traps; PbS QDs as NIR sensitizers.	61	[15]
2015	PDTP-DFBT:ZnO (Vertical)		300–900	≈1000	–	5.1×10^{10} (−4.5 V)	The effect of the number and the size of NPs on the performance of PM phenomenon	141	
2016	PVB:Y-TiOPc (Vertical)		NIR (780 nm)	3 54 200	–	3.1×10^{14} (15 V)		142	
2015	P3HT:$PC_{71}BM$(100:1)/LiF (Vertical)		300–700	16 700	≈0.1	–	The first PM type OPDs with acceptor as electron trap	43	
2015	P3HT:PTB7-Th:$PC_{71}BM$ (50:50:1) (Vertical)		300–800	38 000	≈0.1	1.9×10^{13} (−25 V)	Ternary strategy extend spectral response to NIR	63	
2016	P3HT:ITIC (100:1) (Vertical)		300–790	3650 5750	–	7.0×10^{13} (+15 V) 8.3×10^{12} (−15 V)	The OPDs work well at bi-directional bias, and non-fullerene acceptor as trap.	62	[17]
2018	P3HT:PTB7-Th:$PC_{61}BM$ (40:60:1, thick) (Vertical)		350–800 (broadband) λ_m= 800 FWHM≈40 nm	600 200	–	2.9×10^{11} (+10 V) 2.4×10^{10} (−10 V)	The combination of broadband and narrowband detection	87	
2018	P3HT/$PC_{71}BM$ (Vertical)		300–700	1 20 700	–	$>10^{13}$ (−19 V)	PM type OPDs based on PHJ	47	[19]
2017	P3HT:$PC_{71}BM$ (100:1, thick) (Vertical)		λ_m= 650 FWHM<30 nm	53 500	≈0.4	1.3×10^{11} (−10 V)	Narrowband PM type OPDs combining PM with CIN	81	
2016	P3HT:$PC_{61}BM$:CdTe (thick) (Vertical)		λ_m= 660 FWHM>100 nm	200	900 Hz	7.3×10^{11} (−6 V)	Narrowband PM type OPDs based on ternary blend film	79	

(continued)

Table 11.1 (continued)

2016	PDTP-DFBT:$PC_{71}BM$:PbS (thick) (Vertical)		λ m= 890 FWHM < 50 nm	160	1100 Hz	8.0×10^{11} (−7 V)	The hole trap(PbS)-assisted electron tunneling injection	80	[25]
2007	PEDOT:PSS/C60 (Vertical)	In interfacial layer	350–700	5000	–	–	PM induced by charge trapped at interface is much than that by C60 disordered structure.	38	[5]
2014	PEDOT:PSS/C-TPD:ZnO/C60 (Vertical)		300–700	400	–	6.5×10^{12} (−6 V)	Introducing ZnO NPs as traps into C-TPD interfacial layer.	160	
2016	ZnO/PDPP3T:$PC_{71}BM$ (Vertical)	In interfacial layer	300–1000	1 40 000	≈0.27	6.3×10^{12} (−0.5 V)	Oxygen desorption from the surface defects of ZnO NPs reduces the injection barriers.	44	
2015	PbS/TAPC/MoO_3/Ag (Vertical)		700–1500	18 700	≈0.003	7.0×10^{13} (−10 V)	The penetration of Ag atoms as hole trap states into the TAPC interfacial layer	161	
2014	S-Au/P3HT:$PC_{61}BM$ (Vertical)		400–680	1500	–	–	The hole traps at the S-Au interfacial layer	162	[16]
2014	NiO/ZnO (Vertical)		300–400	25 300	6.5	2.0×10^{13} (−1 V)	The NiO film with defects as hole traps	163	
2010	NTCDA/C60/C60:CuPc (Vertical)	Interfacial blocking layer	400–800	34 000	1000 Hz	–	The dual NTCDA/C60 layer as hole blocking layer	42	
2016	4P-NPB/G./ITO/TPBi/C70/SnPc:C70 (Vertical)		200–1000	>10 000	–	4.0×10^{12} (−10 V)	TPBi (or BmPyPB, or LiF)/C70 as hole blocking layer	164	[13]
2016	C60/C60:PbPc (Vertical)		NIR (850 nm)	1952	–	–	The ultrathin C60 film as hole trap or barrier layer	52	
2017	PBDTT-DPP:$PC_{71}BM$/MoO_3 (Vertical)		350–900	≈5000	≈5 × 10^4 Hz	$>10^{14}$ (+1 V)	Photo-induced release of accumulated carriers in MoO_3 interfacial layer	45	

in polymer films was first discovered in 1999 by Daubler and Neher by using a single film of arylamino-PPV (poly(*p*-phenylene vinylene)) [6]. Stimulated by the progress in polymer organic solar cells [7–9], studies on polymer OPM devices mushroomed after the year 2000.

Irrespective of the small molecules and polymers employed in OPM devices, these devices have blended junctions between donor (D) and acceptor (A) molecules. This configuration is obviously strongly influenced by the field of organic solar cells, for which the blended junction has become the standard structure.

From the standpoint of the multiplication mechanism, irrespective of the small molecules and polymers, researchers have focused on the following three issues:

(1) Blended junction
(2) Carrier blocking layer
(3) Carrier trap

11.3 Blended Junction

The photocarrier-generation efficiency of organic semiconductors is drastically enhanced by blending donor (D) and acceptor (A) molecules. The author invented this blended junction (bulk heterojunction) for organic solar cells in 1991 [10, 11]. In Chap. 7 (Sect. 7.3), the author discusses the introduction of a blended junction into the multiplication device [12]. The co-deposited photoactive layer between fullerene and copper phthalocyanine (C_{60}:CuPc) shows a high-speed response. Using the D/A blended junction, the photocarrier generation efficiency of the primary photocurrent increased to 60% due to enhanced exciton dissociation at the D/A interface. Thus, a sufficient number of holes for electron injection quickly accumulated in the interfacial traps at the C_{60}:CuPc/metal interface, resulting in a more rapid response (within 10 ms). A higher carrier generation efficiency of the primary photocurrent is obviously advantageous for enhancing the OPM performance, irrespective of the small molecules and polymer devices. Over the last decade, blended junctions have become the general structure for OPM devices (Table 11.1).

11.4 Blocking Layer

11.4.1 Blocking Layer at Active Interface

In 2016, Yang and Ma reported OPM cells with the ITO/TPBi/C_{70}/TAPC (or SnPc):C_{70}/BCP/Al structure (Fig. 11.1a) [13, 14]. As mentioned in Sect. 11.3, the blended films of TAPC:C_{70} and SnPc:C_{70} fabricated using the co-evaporation technique acted as the photoactive layer. The TPBi film acts as a hole-blocking layer (HBL) (Fig. 11.2a). The multiplication rate reached 10^4% (a 100-fold increase). For

Chapter 12
Perspective on Organic Photocurrent Multiplication

12.1 Aim of This Chapter

The origin of the organic photocurrent multiplication is the carrier trap. The carrier trap was attributed to the molecular steps and kinks acting as a molecular blind alley (dead end). Surprisingly, the macroscopic device function of multiplication rate reaching 10^5-fold and the nanostructure of molecular steps and kinks are directly connected. This indicates that a large area of uncultivated fields directly connects the nanostructure and macroscopic device functions. Therefore, this perspective was first discussed. Additionally, we discussed the role of the insulating layer from the perspective of tunneling and the perspective of organic avalanche multiplication.

The discussion in this chapter will pave the way for future progress in organic photocurrent multiplication research.

12.2 Artificial Fabrication

12.2.1 Electrode Metal Having the Tip of the Needle Shape

The molecular blind alley model is shown in Fig. 12.1a. Our observations revealed that the electrode metal is spherical (Chap. 5). Electron injection occurs at the top of the metal sphere. Because the electric field is concentrated more intensely at the tip of the needle shape (Fig. 12.1b), a more efficient electron injection can be expected. Such a tip array (Fig. 12.1c, d) [1, 2] for field emission can be fabricated by utilizing the fabrication technology for field emission display [3, 4].

M. Hiramoto, *Organic Photocurrent Multiplication*, Electronic Materials: Science & Technology, https://doi.org/10.1007/978-981-99-1237-7_12

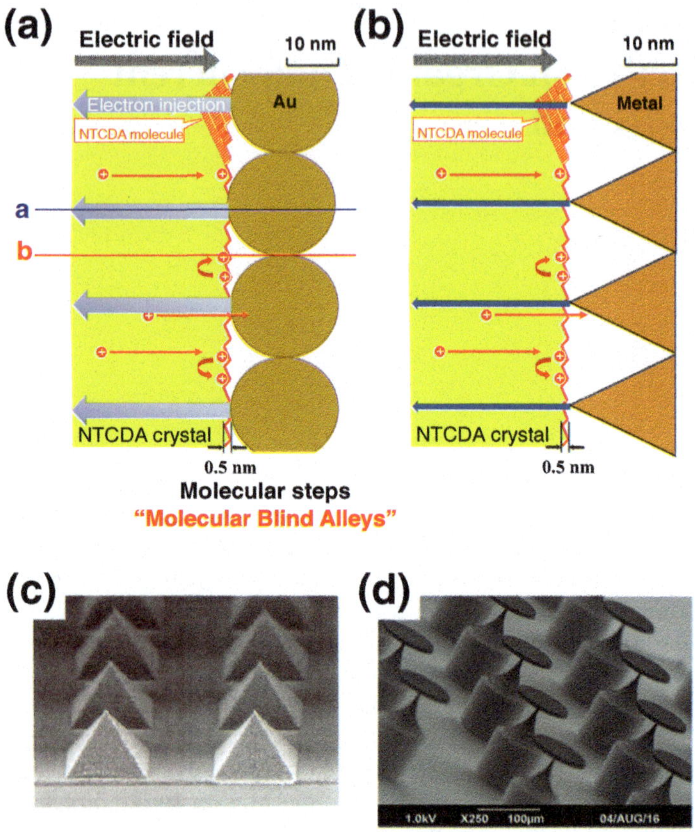

Fig. 12.1 **a** Molecular blind alley model proposed in Chap. 5. The electrode metal had spheric nanoparticle morphology. **b** Advanced molecular blind alley model. Shape The electrode metal had a needle-tip morphology. **c** SEM photograph pyramid (2 μm footprint) array. **d** SEM photograph of commercially available emitter array. Reproduced with the permission of [1]. © Springer-Verlag Berlin Heidelberg 2015

12.2.2 Molecular Blind Alley Traps Made by Nanoimprint Lithography

The depth of the molecular blind alley in Fig. 12.1 is 0.5 nm, i.e., the height of the molecular step of the NTCDA single crystal surface (see Chap. 5). Such molecular-sized blind alleys can be fabricated artificially using nanoimprint lithography (Fig. 12.2b) [1, 4–8]. Nanoimprint lithography has attained sub-3 nm resolution [8]. Rectangular molecular blind alleys (Fig. 12.2a). Depth (d), lateral size (L1), and neighboring distance (L2) were designed. A larger d corresponds to higher energy for releasing trapped holes from the molecular blind alley. This presumably resulted in a slower decay rate in the response profile. A larger L2, and therefore a

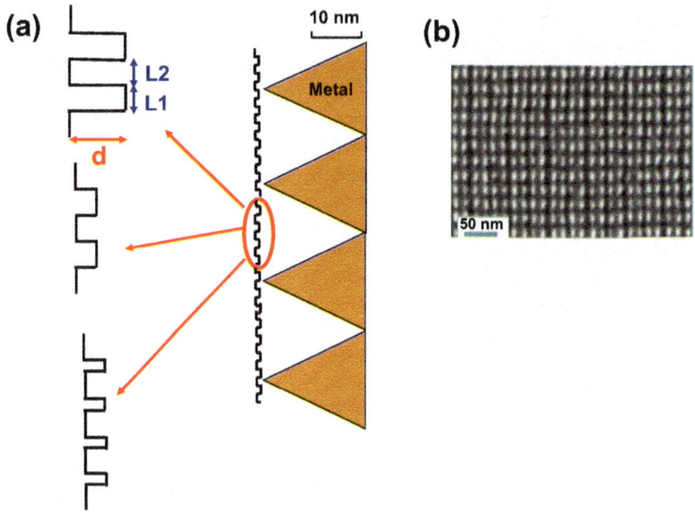

Fig. 12.2 **a** Artificial molecular blind alleys having a rectangular shape. The three left illustrations are enlarged views. Depth (d) increased from the middle left to the upper left. The trap density, increased from the middle left to the lower left. **b** 7 nm half-pitch patterns made by imprint lithography. Reproduced with the permission of [7]. © Springer

decrease in trap density, is expected to result in a lower multiplication rate and faster response. Various shapes of molecular blind alleys, other than rectangular shapes, can be designed. Nanoimprint lithography can be applied to materials, such as inorganic semiconductors, amorphous organic films, and polymers. Therefore, multiplication devices can be fabricated using various types of material.

12.2.3 Micrometer-Sized Thickness Heterogeneity Made by Nanoimprint Lithography

The multiplied photocurrent oscillation was caused by the heterogeneity of the H_2Pc film thickness (see Chap. 9). Without roughness (thickness heterogeneity), no oscillation occurred (Fig. 12.3a, c). With a thickness heterogeneity of approximately 20 nm and a lateral length of approximately 1 μm, multiplied photocurrent oscillation occurs (Figs. 12.3b). Nanoimprint lithography can easily fabricate such micrometer-size heterogeneity in the lateral direction to the H_2Pc film surface (Fig. 12.3d). Therefore, nanoimprint lithography can be applied to control multiple photocurrent oscillation behaviors.

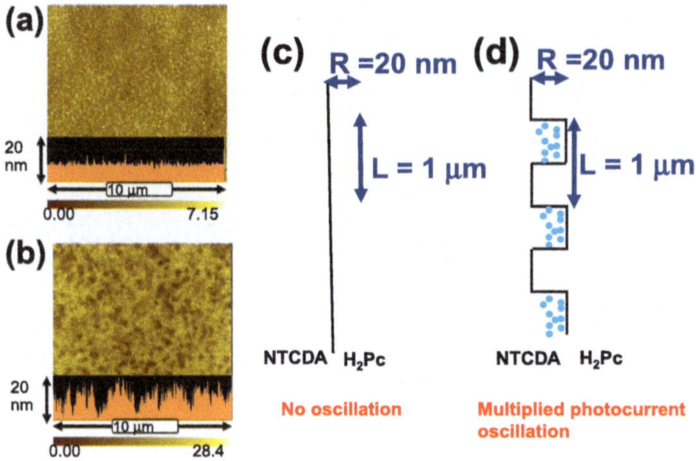

Fig. 12.3 AFM images of H_2Pc film without (**a**) and with roughness (**b**) showing no oscillation and multiplied photocurrent oscillation, respectively. Schematic illustrations of NTCDA/H_2Pc interfaces without (**c**) and with roughness (**d**)

12.3 Direct Design of Molecular Blind Alley

12.3.1 Detailed Observation of Steps and Kinks

The carrier trap causes the multiplication of the organic photocurrent. In Chap. 5, the molecular kinks formed during the molecular steps are identified as molecular blind alleys that act as carrier traps. Therefore, a detailed investigation of the steps and kinks on the organic crystal surface should be performed.

Figure 12.4 shows vivid images obtained by the high spatial resolved scanning probe microscope (SPM). Figure 12.4a shows an atomic force microscopy (AFM) image of the surface of a single crystal of pentacene [9]. The detailed nanostructure of the resolved molecular steps is complex. It is winding, and there are many prickles and a peninsula. Figure 5.19b shows a scanning tunneling microscopy (STM) image of monolayer NTCDA on an Ag(110) surface [10]. Images of NTCDA molecules were vividly resolved. Molecular-level images of the steps and kinks can be distinctly observed. Molecular-level kink structures should be clarified by magnifying the step structure and their relationship with multiplication behaviors.

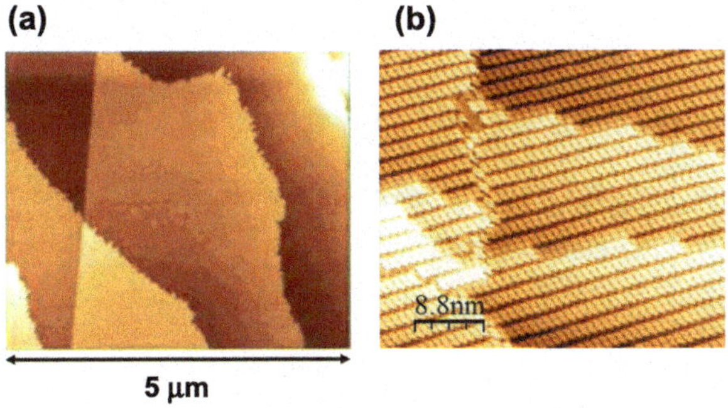

Fig. 12.4 **a** AFM image of a pentacene single crystal [9]. The detailed nanostructure of molecular steps is resolved. **b** STM image of monolayer NTCDA on Ag(110) surface [10]. Molecular-level observation of steps and kinks are achieved. Reproduced with the permission of [9] and [10]. © 2018 American Chemical Society

12.3.2 Control of Multiplication Behaviors by Designing Steps and Kinks

Multiplication behaviors are expected to be controlled by the direct design of molecular steps and kinks that act as molecular blind alleys. The depth and density of traps are crucial factors for controlling multiplication behaviors.

(1) **Trap depth**: A longer/shorter spatial depth of a molecular blind alley can be formed using organic molecules of a larger/smaller size. This directly changes the energetic trap depth as the holes move against the concentrated electric field to escape from the molecular blind alley (see Fig. 12.1). Relationships between multiplication behaviors and highly resolved structures of molecular steps and kinks should be investigated for several typed of organic semiconductor crystals.
(2) **Trap density**: The density of steps and kinks can be controlled by the following methods:

  (i) Various crystal facets by cutting
  (ii) Crystal growth conditions
  (iii) Homoepitaxial growth

  Control of the step and kink structures by utilizing homoepitaxial growth on the crystal surface under various substrate temperatures (see Fig. 5.17 in Chap. 5). Because homoepitaxial growth is closely related to the dynamic growth of steps and kinks, this method is a powerful tool.

  (iv) Thermal treatment

  Rearrangement of step and kink structures by the thermal treatment of crystals.

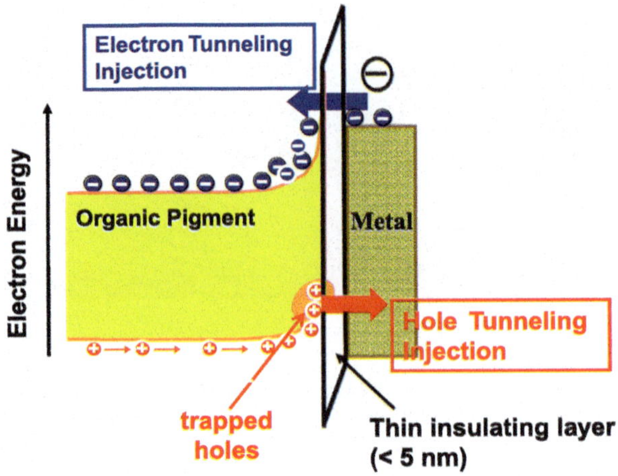

Fig. 12.5 Energy structure of metal/insulator/semiconductor (MIS) junction. The thickness of insulator layer was less than 5 nm. Tunneling injection of electrons and holes occurred equally. In this case, no photocurrent multiplication occurred

12.4 Insulating Layer

12.4.1 Metal/Insulator/Semiconductor (MIS) Junction

For charge injection-type photocurrent multiplication, a very thin insulating layer is often inserted between the electrode metal and semiconductor, and a metal/insulator/semiconductor (MIS) junction is formed (Fig. 12.5). When the thickness of the insulator layer is less than 5 nm, tunneling injection of electrons from the metal electrode to the conduction band of the semiconductor occurs (Fig. 12.5, blue arrow). However, simultaneous tunneling injection of holes from the valence band to the metal electrode occurred (Fig. 12.5, red arrow). Even if holes are trapped at the insulator/semiconductor interface, tunneling injection of holes from such traps to the metal electrode still occurs. Because electrons and holes can tunnel thin insulators equally, trapped holes cannot accumulate under this condition.

12.4.2 Spatial Separation

In the molecular blind alley model, there are sites of direct (Fig. 12.1a, blue line a) and indirect (Fig. 12.1a, red line b) contacts between the Au sphere and NTCDA crystal surface. At indirect contact sites, we can assume that a relatively thick insulator layer (vacuum), which can prevent tunneling, is inserted between the Au and NTCDA. The different energetic structures for these two types of contacts are illustrated (Fig. 12.6). At the direct-contact site (Fig. 12.6a), the tunneling injection of

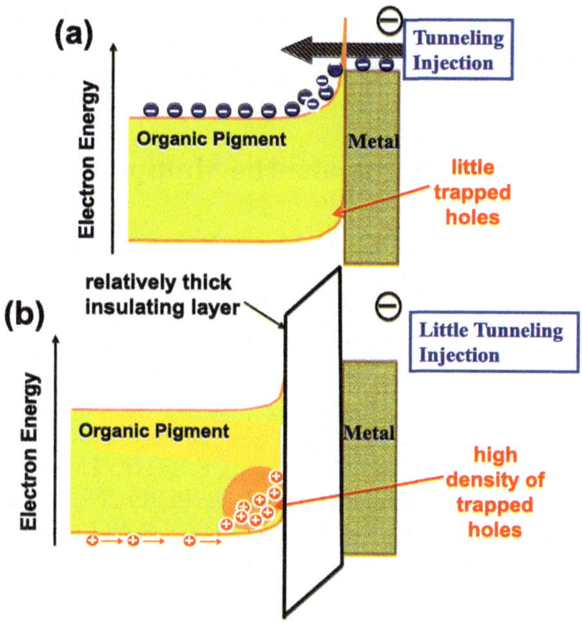

Fig. 12.6 Energy structures based on the molecular blind alley model (Fig. 12.1). **a** At direct contact sites (Fig. 12.1, blue line a). **b** At indirect contact sites (Fig. 12.1, red line b)

electrons predominantly occurs under small trapped holes. At the indirect contact site (Fig. 12.6b), a high density of holes was trapped at the organic surface neighboring the insulating layer, and a small tunneling injection of electrons occurred.

Because the sites of hole trapping and electron injection are spatially separated, the tunneling balance between electrons and holes is broken. Holes are trapped owing to the thick insulating layer, which prevents tunneling. Electrons are injected by tunneling through a very thin Schottky barrier owing to the concentrated electric field. Spatial separation is the key to breaking the tunneling balance between the electrons and holes.

12.4.3 How to Break the Tunneling Balance Between Electrons and Holes?

Generally, to accumulate holes and induce photocurrent multiplication, the tunneling balance between the electrons and holes should be broken. The unknown method used to break the tunneling balance between electrons and holes is a common issue for any device that utilizes charge-injection-type photocurrent multiplication.

Most studies on organic photocurrent multiplication reported after 2000, described in Chap. 11, do not explicitly describe the methodology of breaking the tunneling balance between electrons and holes. However, this is an essential point related to the origin of photocurrent multiplication. A detailed study on this issue should be therefore be performed.

12.5 Organic Avalanche Multiplication

In Chap. 10, avalanche multiplication of the photocurrent in perylene molecular crystals with a band-conductive nature is discussed. This phenomenon is a type of multiplication based on a different mechanism compared to that based on the interfacial tunneling injection described in other chapters of this book. Organic avalanche multiplication provides a more advanced function than organic photocurrent multiplication.

In the last decade, several types of band-conductive organic semiconductors have been reported (Fig. 12.7) [11]. C8-BTBT, TMTSF, rubrene, and TIPS-pentacene exhibited band conduction for holes. PDIF-CN_2 shows band conduction for electrons. The hole mobilities of rubrene and C8-BTBT reached 40 [12] and 10 $cm^2\ V^{-1}\ s^{-1}$ [13, 14].

Organic crystals, such as rubrene single crystals (Fig. 12.8a) and C8-BTBT, could exhibit avalanche multiplication. Figure 12.8b shows a schematic illustration of the measurements [15]. High mobility appeared along the lateral direction to the crystal surface, that is, in the same direction as π stacking. The distance between the surface electrodes should be less than 1 μm to apply a high electric field above 1×10^6 $V\ cm^{-1}$ (Chap. 10). A photocarrier was generated at the donor (rubrene or C8-BTBT)/acceptor (C_{60}) interface. The photogenerated holes accelerated along the band-conductive π -stacking. In the case of avalanche multiplication, an increase in the multiplication rate was observed when the electrode distance was increased by changing the location of the electrode from A to C (Fig. 12.8b).

For organic avalanche multiplication, a faster response reaching the nanosecond order would be obtained compared to that of the microsecond order for charge-injection-type multiplication. This is because the accumulation of trapped carriers requires time (see Chap. 11, Fig. 11.9).

We believe that we can develop avalanche photodetectors, fabricated using organic molecular crystals and films, that show a band-conductive nature.

12.6 Conclusion

Designated fabrication of molecular blind alleys is possible through nanoimprint lithography. Moreover, the direct design of molecular steps and kinks, which act as molecular blind alleys, would be possible. The trap depth can be controlled by controlling the size of the organic molecules. The trap density, that is, the density of kinks, was controlled by the exposure of various crystal facets by cutting, crystal

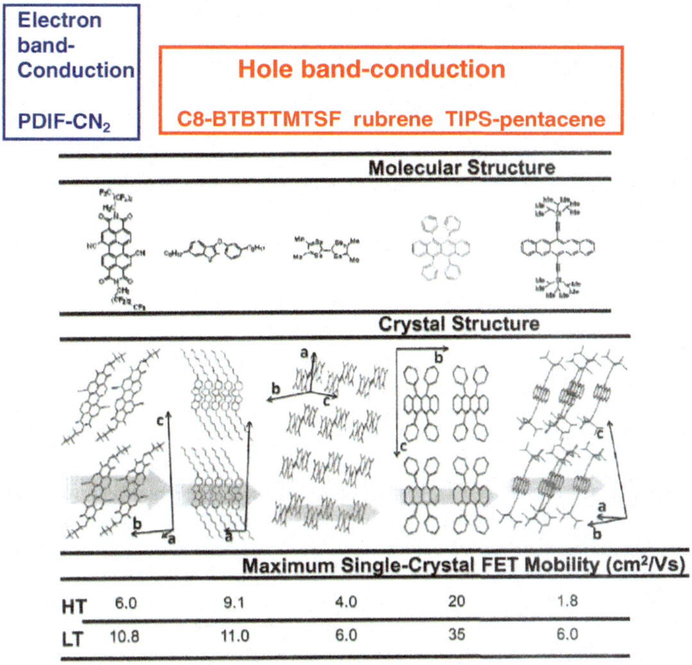

Fig. 12.7 Molecular structure and packing for PDIF-CN_2, C8-BTBT, TMTSF, rubrene, and TIPS-pentacene, which exhibit band-like transport through π-stacking (arrows in the crystal structures). The values of mobility at room temperature (HT) or low temperature (LT) are shown. Reproduced with the permission of [11]. © 2012 WILEY–VCH Verlag GmbH & Co. KGaA, Weinheim

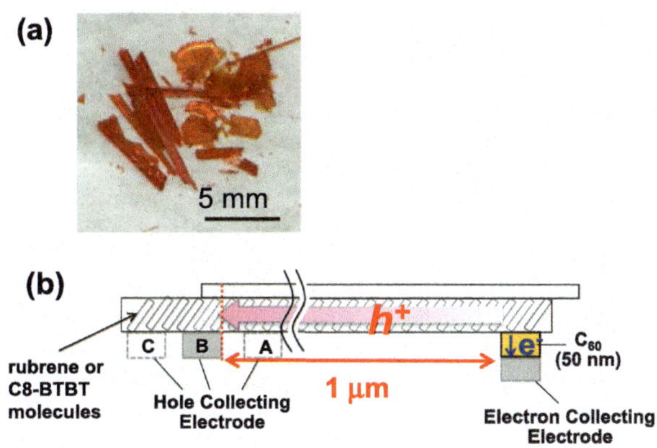

Fig. 12.8 **a** Photograph of single rubrene crystals. **b** Cell structure for organic avalanche multiplication measurements

growth conditions, homoepitaxial growth, and thermal treatment. For the insulating layer, tunneling imbalance through the insulating layer between electrons and holes should be attained to induce organic photocurrent multiplication. Finally, the possibility of organic avalanche multiplication using organic band-conductive semiconductors was proposed.

References

1. Schift, H.: Nanoimprint lithography: 2D or not 2D? A review. Appl. Phys. A **121**, 415–435 (2015)
2. https://www.samco.co.jp/products/process/silicon/bosch-process-drie/Microfabricated_emitter_array_for_an_ionic_liquid_electrospray_thruster.php
3. Spindt, C.A.: A thin-film field-emission cathode. J. Appl. Phys. **39**, 3504 (1968)
4. Spindt, C.A., Holland, C.E., Rosengreen, A., Brodie, I.: Field-emitter arrays for vacuum microelectronics. IEEE Trans. Electron Devices **38**, 2355–2363 (1991)
5. Chou, S.Y., Krauss, P.R., Renstrom, P.J.: Imprint of sub25 nm vias and trenches in polymers. Appl. Phys. Lett. **67**, 3114–3116 (1995)
6. Hirai, Y.: Polymer science in nanoimprint lithography. J. Photopolym. Sci. Tech. **18**, 551–558 (2005)
7. Sreenivasan, S.V.: Nanoimprint lithography steppers for volume fabrication of leading-edge semiconductor integrated circuits. Microsyst. Nanoeng. **3**, 17075 (19 pp) (2017)
8. Hua, F., Sun, Y., Gaur, A., et al.: Polymer imprint lithography with molecular-scale resolution. Nano Lett. **4**, 2467–2471 (2004)
9. Sato, K., Sawaguchi, T., Sakata, M., Itaya, K.: Noncontact atomic force microscopy of perfect single-crystals of pentacene prepared by crystallization from solution. Langmuir **23**, 12788–12790 (2007)
10. Tong, Y., Fuhr, J.D., Martiarena, M.L., Oughaddou, H., Enriquez, H., Nicolas, F., Chaouchi, K., Kubsky, S., Bendounan, A.: Properties of NTCDA thin films on Ag(110): scanning tunneling microscopy, photoemission, near-edge X-ray fine structure, and density functional theory investigations. J. Phys. Chem. C **123**, 379–386 (2019)
11. Minder, N.A., Ono, A., Chen, Z., Facchetti, A., Morpurgo, A.F.: Band-like electron transport in organic transistors and implication of the molecular structure for performance optimization. Adv. Mater. **24**, 503–508 (2012)
12. Takeya, J., Yamagishi, M., Tominari, Y., Hirahara, R., Nakazawa, Y.: Very high-mobility organic single-crystal transistors with in-crystal induction channels. Appl. Phys. Lett. **90**, 102120 (3 pp) (2007)
13. Yuan, Y., Giri, G., Ayzner, A.L., Zoombelt, A.P., Mannsfeld, S.C.B., Chen, J., Nordlund, D., Toney, M.F., Huang, J., Bao, Z.: Ultrahigh mobility transparent organic thin film transistors grown by an off-centre spin-coating method. Nat. Commun. **5**, 3005 (5 pp) (2014)
14. Haas, S., Takahashi, Y., Takimiya, K., Hasegawa, T.: High performance dinaphtho-thieno-thiophene single crystal field-effect transistors. Appl. Phys. Lett. **95**, 022111 (2009)
15. Kikuchi, M., Hirota, M., Kunawong, T., Shinmura, Y., Abe, M., Sadamitsu, Y., Moh, A.M., Izawa, S., Izaki, M., Naito, H., Hiramoto, M.: Lateral alternating donor/acceptor multilayered junction for organic solar cells. ACS Appl. Energy Mater. **2**, 2087–2093 (2019)

The manufacturer's authorised representative in the EU is Springer Nature Customer Service Centre GmbH, Europaplatz 3, 69115 Heidelberg, Germany. If you have any concerns regarding our products, please contact ProductSafety@springernature.com

Printed and bound by CPI Group (UK) Ltd, Croydon, CR0 4YY
30/09/2026
02225308-0001

the SnPc:C_{70} cell, the response wavelength region extended from the near-infrared region (NIR) to 1000 nm (Fig. 11.2b). They concluded that a large number of holes accumulated in the C_{70} layer because the photogenerated holes from the blended photocurrent generation layer are blocked by TPBi, which acts as a hole-blocking layer. Thus, the applied voltage is concentrated in the TBPi layer. In addition, the width of the barrier to electron tunneling from the ITO electrode was significantly reduced because of band bending (Fig. 11.1b). Multiplication is caused by electron tunneling, which in turn is caused by hole accumulation at TPBi/C_{70}. These observations clearly indicate that inserting an interfacial blocking layer can induce the accumulation of holes and trigger the multiplication process.

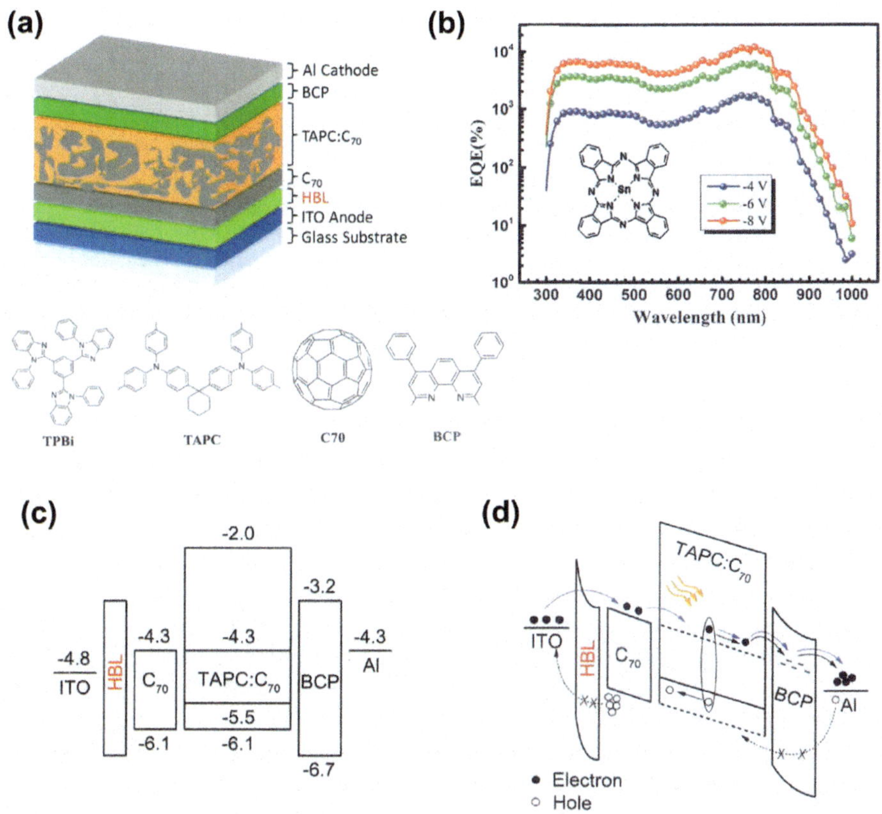

Fig. 11.1 **a** Device structure of organic photocurrent multiplication (OPM) device. The hole blocking layer (HBL) is TPBi. **b** EQE spectra of device with ITO/TPBi (3 nm)/C_{70} (10 nm)/SnPc:C_{70} (70 nm)/BCP (10 nm)/Al structure. **c** Schematic energy diagram of the device. **d** Photomultiplication mechanism. Reproduced with the permission of [14] and [13]. © 2020, © 2016 The Royal Society of Chemistry

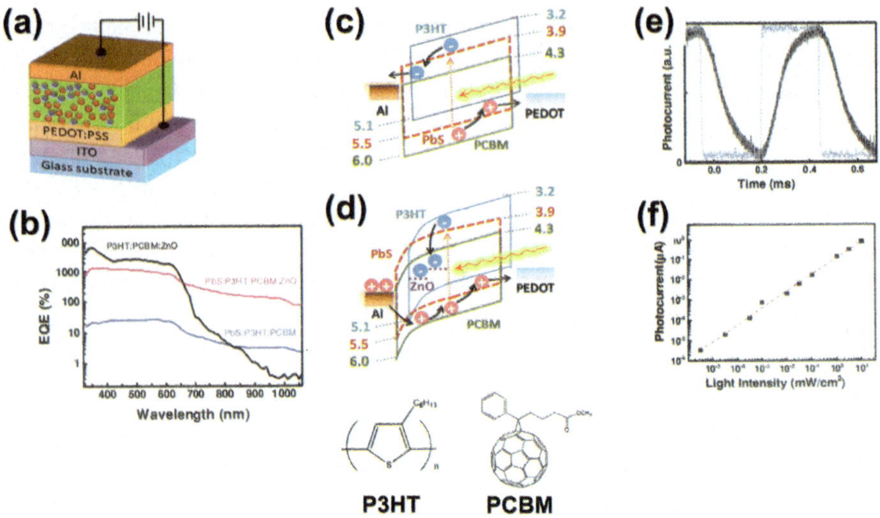

Fig. 11.2 **a** Device structure with uniform traps. ZnO QDs acting as electron traps are added to the PbS:P3HT:PCBM ternary blend. **b** Spectral dependence of external quantum efficiency (EQE). **c** Energy diagram of PbS:P3HT:PCBM ternary blend under operation. **d** Energy diagram of ZnO QD-doped PbS:P3HT:PCBM ternary blend under operation. **e** Multiplied photocurrent response under a bias of −3 V at a light intensity of 0.1 mW/cm^2 (black curve). **f** Photocurrent versus input NIR light (808 nm). Chemical structures of P3HT and PCBM are shown in the figure. Reproduced with the permission of [15]. © 2014, WILEY–VCH Verlag GmbH

11.4.2 Blocking Layer at Counter Interface

The insertion of a blocking layer at the counter electrode interface, that is, the electrode interface opposite to that at which multiplication occurs, provides an additional effect. For example, a wide-bandgap electron-transporting molecule, BCP, was inserted between the aluminum electrode and the active blended layer to prevent the injection of holes from the Al anode (Fig. 11.2a) [13, 14]. If holes can be injected into the active layer from the positively biased Al electrode, the injected holes can trigger the multiplication process, even under dark conditions. Therefore, inserting the blocking layer at the counter electrode side is very effective in suppressing the dark current. Moreover, because this device has blocking layers at both electrode interfaces, it is difficult to inject both electrons and holes into the active layer under dark conditions. Therefore, further suppression of the dark current is achieved.

For photosensors, dark current suppression is crucial for detecting very weak light. That study showed that the insertion of a blocking layer at both interfaces is very effective for constructing practical OPM devices.

11.5 Carrier Trap

Carrier traps are the primary origin of photocurrent multiplication. Therefore, many studies on the intentional formation of various types of carrier traps have been performed.

11.5.1 Uniform Trap

As a typical example, we describe intentional doping of films with ZnO quantum dots (QDs), which act as uniform traps [15]. Figure 11.2a shows the cell structure. The active layer is a polymer-blended layer of P3HT:PCBM, which is a benchmark system for polymer organic solar cells [9]. PbS QDs, acting as near-infrared (NIR) sensitizers, were added. The ZnO QDs acted as electron traps and were uniformly incorporated into the blended film.

Figure 11.2b shows the spectral dependence of the external quantum efficiency (EQE) under a large reverse bias. Although the PbS:P3HT:PCBM ternary blended device exhibited a NIR response, its EQE was smaller than 100% (violet curve). However, by doping with ZnO QDs, the EQE exceeded 100%, approaching a value near 1,000% (10-fold) (red curve). The P3HT:PCBM blended film with ZnO QDs also exhibited an EQE value near 10,000% (100-fold) (black curve).

Figure 11.2c shows the energy diagram of the PbS:P3HT:PCBM ternary blended device. The photogenerated electrons and holes in the PbS QDs were transferred to PCBM and P3HT. Because of the large injection barrier for holes between the Al electrode and the highest occupied molecular orbital (HOMO) of the PbS QDs, P3HT, and PCBM, the ternary blend film devices showed no multiplication. In contrast, as shown in Fig. 11.2d, for the device with doped ZnO QDs, the trapped electrons in the ZnO QDs induce band-bending close to the positively biased interface and trigger hole injection under reverse bias, resulting in photocurrent multiplication.

This device showed a very fast response to the multiplied photocurrent (Fig. 11.2e, black curve). A rise time (change in the output signal from 10 to 90% of the peak output value) of 160 μs and a decay time of approximately 80 μs after switching off the NIR pulse (violet curve) were observed. This device also exhibited a very large dynamic range of light detection, that is, the photocurrent was linearly proportional to the input NIR light within the range of 10 mW/cm^2–3.1 nW/cm^2 (Fig. 11.2f).

11.5.2 Interfacial Traps

As a typical example of an interfacial trap, we describe semi-continuous Au (s–Au) acting as hole traps at the ITO electrode/blended polymer (PCBM:P3HT (1:1)) interface [16]. Figure 11.3a shows the device structure. The s–Au layers were sputtered

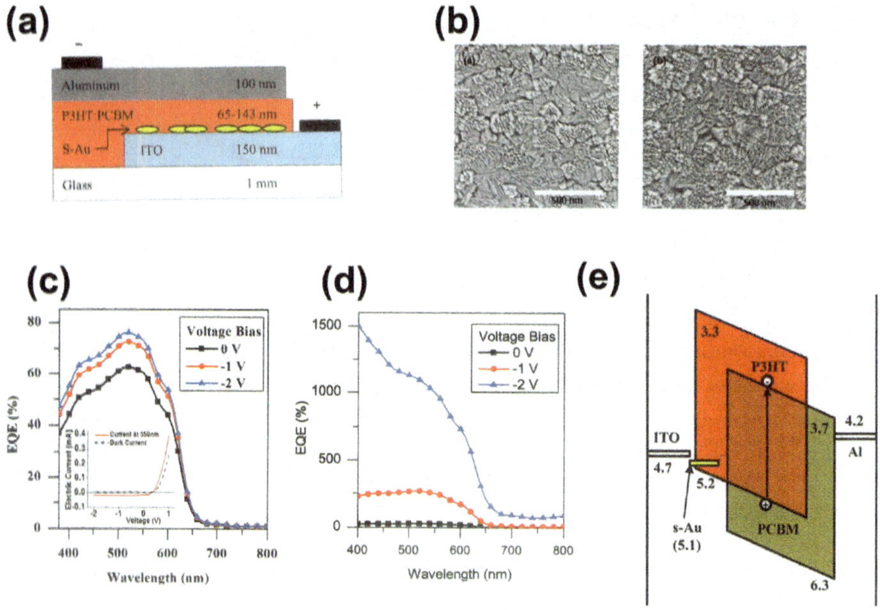

Fig. 11.3 **a** Device structure with interfacial traps. Semi-continuous Au (s–Au) acting as hole traps are inserted at the ITO/polymer interface. **b** SEM images of the ITO surfaces without (left) and with (right) s–Au. The sharp ITO nanocrystals are nucleation sites for the formation of the s–Au layer. The scale bar is 500 nm. **c** EQE of device without s–Au. **d** EQE of device with s–Au. **e** Energy structure of the device. Reproduced with the permission of [16]. © 2014, AIP Publishing LLC

directly onto ITO substrates. Figure 11.3b shows the SEM images of ITO with and without Au. Au grew on the sharp surfaces of the ITO crystals.

Figure 11.3c shows the external quantum efficiency (EQE) of a device with a 65 nm thick P3HT:PCBM active layer, without s–Au. This device exhibited no multiplication, that is, the EQE remained below 80%. Figure 11.3d shows the EQE response of a device with the same active layer thickness as the s–Au film at the ITO/P3HT:PCBM interface. This device showed an extreme enhancement of the EQE response across the entire range of wavelengths tested. At 1 V, the EQE increased sharply, that is, by more than 200% (2-fold) from 400 to 600 nm. At 2 V, the EQE increased sharply and approached 1,500% (15-fold) at 400 nm.

In this case, the s–Au film acts as a hole-blocking layer or hole-trapping site (Fig. 11.3e). As the positive charges build at the s–Au/P3HT:PCBM interface, an intense electric field forms between the s–Au and ITO anode, resulting in the injection of multiple electrons into the device.

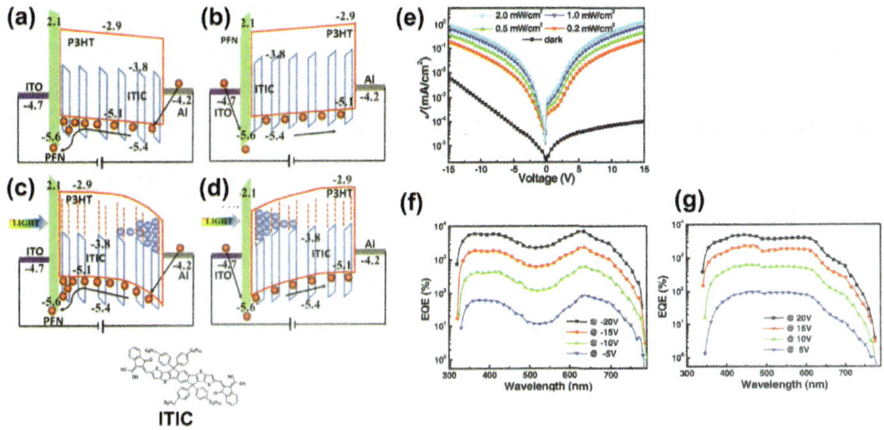

Fig. 11.4 a–d Energy structures of the device under dark and reverse bias (**a**), under dark and forward bias (**b**), under light irradiation and reverse bias (**c**), and under light irradiation and forward bias (**d**). **e** Current density–voltage (*J–V*) curves from −15 to 15 V under dark conditions and under white-light illumination at several different intensities. **f** EQE spectra under different reverse biases. **g** EQE spectra under different forward biases. Reproduced with the permission of [17]. © 2016, WILEY–VCH Verlag GmbH

11.5.3 Incomplete Percolation Traps

When the ratio of donor (D) or acceptor (A) in the D/A blended film is very low, an incomplete percolation end and localized islands are formed, which can act as carrier traps. Here, we call this type of trap an "incomplete percolation trap". For example, when polymer blends of P3HT (donor) and ITIC (acceptor) with a weight ratio of 100:1 were used, the ITIC acceptor formed incomplete percolation traps [17]. ITIC is a non-fullerene acceptor (NFA) (Fig. 11.4) [18].

The working mechanism of this device under both reverse and forward bias is shown for both dark and light conditions (Fig. 11.4a–d). In the dark, a large interfacial barrier prevents hole injection under both reverse and forward biases (Fig. 11.4a, b). Under light irradiation, trapped electrons in the incomplete percolation traps in the ITIC region near the Al electrode (Fig. 11.4c) or ITO electrode (Fig. 11.4d) trigger hole tunneling injection for the reverse and forward bias, respectively. The majority of photogenerated electrons are directly trapped in the ITIC near the ITO electrode, resulting in P3HT band-bending at the interface.

This device shows a rather low dark current under forward and reverse biases (Fig. 11.4e). Moreover, the photocurrent is super-linearly enhanced with the applied bias and with an increase in the incident light intensity. From the EQE spectra obtained under reverse and forward biases (Fig. 11.4f, g), the EQE was close to 10^4% (100-fold). This means that more holes were injected from the ITO and Al electrodes into the P3HT:ITIC active layer. Thus, photocurrent multiplication occurred for both bias polarities.

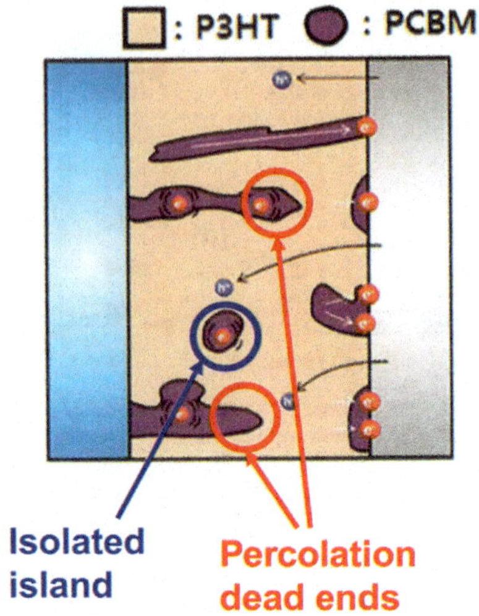

Fig. 11.5 Schematic of incomplete percolation traps. Isolated island (blue circle) and percolation dead ends (red circles) are shown. Reproduced with the permission of [19]. © 2018, American Chemical Society

Figure 11.5 presents a schematic of incomplete percolation traps [19]. Here, the regions of the donor (P3HT) and acceptor (PCBM) molecules are color-coded orange and violet, respectively. The PCBM can be replaced with an ITIC. When the acceptor (PCBM) ratio is low in the blended film (Fig. 11.5a), the route for generation of electrons becomes incomplete, and isolated islands (blue circles) and percolation dead ends (red circles) are formed. These structures act as electron traps. In the case of localized islands, the electrons formed at the donor/acceptor interface are automatically confined to the localized island, that is, they are captured by the trap. In the case of percolation dead ends, when the right-side electrode is positively biased with respect to the electrode on the left side, the percolation dead ends can act as electron traps because the electrons move toward the dead end. Notably, the percolation dead end seems to have some features in common with the structural trap model proposed by the author in Chaps. 4 and 5 in this book.

As another example, isolated islands formed at the interface act as electron traps (Fig. 11.6). These traps are a type of incomplete percolation trap, and have been reported to cause multiplication [19].

11.6 Universality of Multiplication Concept—Hybrid System

Evaluation of a hybrid system consisting of various organic and inorganic materials revealed the possibility of constructing efficient multiplication systems. Ishii et al. reported that the molecular layer of an organo-europium complex (Eu-terpy)

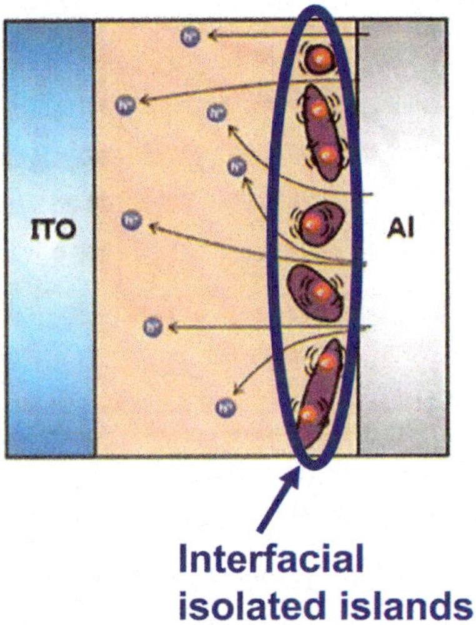

Fig. 11.6 Schematic of interfacial isolated islands acting as electron traps. Reproduced with the permission of [19]. © 2018, American Chemical Society

hybridized with the surface of halide perovskite ($MAPbI_3$) nanocrystals caused efficient photomultiplication (Fig. 11.7a) [20, 21]. Hole accumulation in the photoexcited perovskite nanoparticles induced electron tunneling across the Eu- complex molecules.

The EQE action spectrum in Fig. 11.7b shows the multiplied photocurrent under the application of a low external bias of −0.5 V. An EQE value of approximately $2.5 \times 10^5\%$, that is, a multiplication rate of 2,500-fold was observed from 500 to 800 nm, which matches the optical absorption of the halide perovskite ($MAPbI_3$). Without the Eu complexes, no multiplication occurred (Fig. 11.7a, yellow curve in *J*–*V* plot). With the Eu complexes, the photocurrent increased drastically from 10^{-6} to 1 A cm^{-2} (Fig. 11.7a, yellow and red curves in *J*–*V* plot); that is, photocurrent multiplication was induced. The signal-to-noise (*S/N*) ratio for photodetection was of the order of 10^6 (compare dark current (black curve). A photocurrent (red curve in *J*–*V* plot in Fig. 11.7a) resulting from strong suppression of the background dark current by the organic ligands (terpy) of the Eu-complex was observed.

Figure 11.7c shows the time-course of the photocurrent response after irradiation with weak monochromatic light (0.76 mW cm^{-2}; 550 nm) under an applied voltage of −0.5 V. Although the multiplied photocurrent density reached ~1 A cm^{-2}, minor transient signals were observed in response to intermittent irradiation with light (on and off) during the timescale of seconds.

A high multiplication rate, fast response, and low signal-to-noise ratio were achieved using this system. This system uses completely different materials than conventional multiplication systems (Table 11.1), proving that the charge injection-type multiplication concept is universal, irrespective of the type of material.

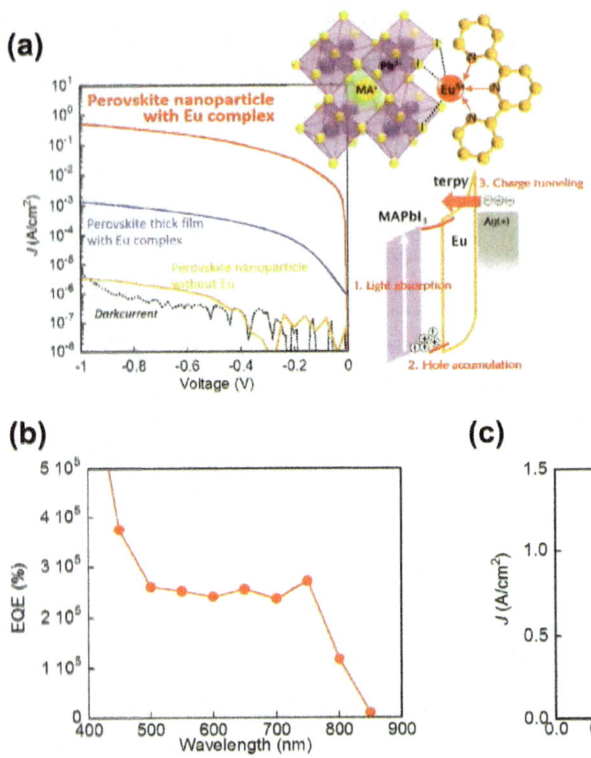

Fig. 11.7 **a** Upper right: Schematic of molecular layer of an organo-europium (Eu) complex (Eu-terpy) hybridized with the surface of halide perovskite ($MAPbI_3$) nanocrystals. Lower right: Energy structure during multiplication process. Left: *J*–*V* characteristics. **b** EQE action spectrum of multiplied photocurrent at −0.5 V. **c** Time-course of photocurrent response under irradiation with 550 nm monochromatic light (0.76 mW cm^{-2}) at −0.5 V. Reproduced with the permission of [21]. © 2019, American Chemical Society

11.7 Multiplication in Charge Transfer Region

For organic solar cells with a blended junction, the charge transfer (CT) state between donor and acceptor molecules is a key issue for photocurrent generation by exciton dissociation, as well as photovoltage generation related to carrier recombination [22, 23]. Therefore, EQE spectra in the CT absorption region have been intensively studied. Recently, Kublitski and Leo reported that multiplication occurs even in the CT absorption region [24]. Device operation (Fig. 11.8a, b) is based on the interfacial blocking layer (see Sect. 11.4) and electron traps due to incomplete percolation, as mentioned in Sect. 11.5.3. Hatna-Cl_6 (Fig. 11.8) was used as a blocking layer. Electron traps were formed by incomplete percolation in a co-deposited film of ZnPc:C_{60} with a small C_{60} ratio of 3 wt%. Owing to the low ratio of C_{60}, C_{60} percolation

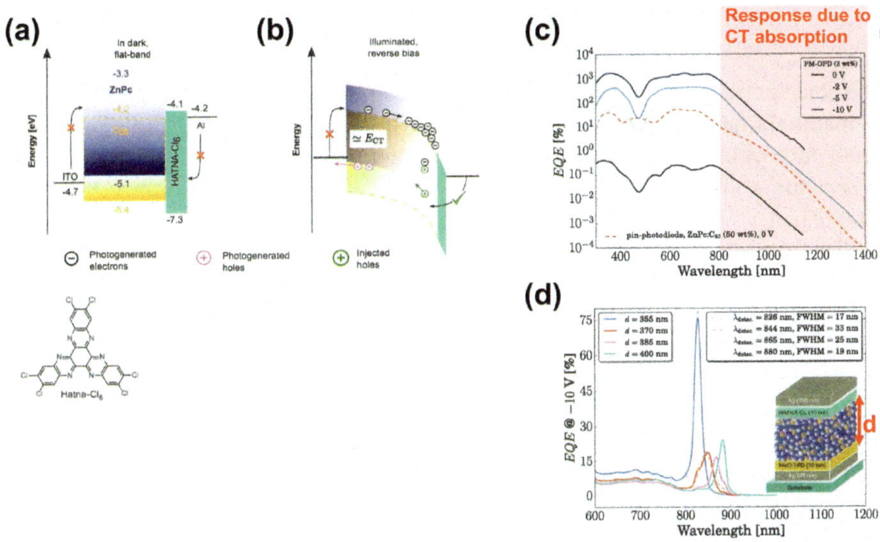

Fig. 11.8 **a** Energy diagram of ITO/ZnPc:C_{60} (C_{60}:3 wt%)/Hatna-Cl_6/Al device in the dark at flat band condition. **b** Energy diagram under illumination and reverse bias. **c** Voltage-dependent EQE spectra. **d** EQE of devices with optical cavities with various active layer thicknesses (**d**). Reproduced with the permission of [24]. © 2021, Springer Nature

dead-end-acting electron traps (Fig. 11.5) were formed. Hole injection was triggered by electron accumulation near the positively biased electrode (Fig. 11.8b).

Figure 11.8c shows the EQE spectra for the device under bias voltages of 0 V (lower black curve), −2 V, (pale blue curve), −5 V (blue curve), and −10 V (upper black curve). Because the onset of the longer wavelength side of the absorption spectrum of ZnPc is approximately 800 nm, the EQE response beyond 800 nm (transparent pink) is attributed to the absorption of the CT state formed between ZnPc and C_{60}. The EQE values were plotted on a logarithmic scale. From 800 to 900 nm, the EQE values were between 10^3 and 10^2%, corresponding to respective multiplication rates of 10 and onefold. Thus, it can be concluded that this multiplication is caused by the absorption of the CT state.

By sandwiching the active blended layer between two Ag electrodes (Fig. 11.8d, inset), an optical microcavity was formed, and narrowband detection of NIR light at specific wavelengths was achieved. By changing the active layer thickness (d), the detection wavelength with a small full width at half maximum (FWHM) was effectively tuned, as shown by the four peaks in Fig. 11.8d.

Because the CT state appears in the longer wavelength region, including the near infrared (NIR) region, the absorption of the CT state enabled NIR detection.

11.8 Essence of Progress

The concept of carrier injection-type multiplication has been verified for many types of materials, such as single-and D/A-blended vacuum-deposited films of organic small molecules, single and D/A blended polymer films, and inorganic–organic hybrid films. Thus, the field of organic photocurrent multiplication (OPM) was established.

The primary origin of OPM is carrier traps. The intentional formation of carrier traps is the main approach for harnessing this phenomenon. The induction of multiplication by intentional doping with uniform traps (Sect. 11.5.1) and interfacial traps (Sect. 11.5.2) was achieved. Incomplete percolation traps, namely percolation dead ends and isolated islands (Sect. 11.5.3), are new trap concepts.

The secondary origin of OPM is the blocking layer. The main approach for manipulating the blocking layer is by intentional insertion of the blocking layer. The blocking layer provides a field of carrier tunneling at the active interface.

11.9 Organic Photocurrent Multiplication (OPM) Devices

From the standpoint of realizing an OPM photodetector, the issues outlined below have been studied. The response speed of the multiplied photocurrent remains an issue.

(i) Reduction of dark current and higher *S/N* ratio: Insertion of the carrier blocking layer at the active interface, as well as at the counter electrode, effectively suppresses the dark current and increases the signal-to-noise (*S/N*) ratio [13, 14]. This allows linearity over a wide light- intensity range [15, 25].

(ii) Response speed: The response speed is still an issue in OPM devices. Figure 11.9 shows the relationship between the EQE and bandwidth, which can be regarded as the reciprocal value of the response speed (red horizontal axis) for OPM devices reported to date [1]. There is a trade-off relationship between the EQE and response speed; that is, a device with a larger EQE has a slower response speed. The response speed of the OPM devices from several seconds to several micro-seconds is still slow compared to that of the conventional photodiode, which reaches a response speed of 0.1 ns [1].

(iii) Near infrared (NIR) detection: To fabricate OPM devices with NIR sensitivity, NIR absorbers such as SnPc (Sect. 11.4.1) [13, 14], PbS (Sect. 11.5.1) [15, 25], and the CT state (Sect. 11.7) [24] have been utilized.

(iv) Narrowband detection of specific wavelengths: The optical microcavity effect (Sect. 11.7) [24] and the light penetration effect at the absorption spectral edge, that is, the masking effect [25], were utilized to achieve narrowband detection.

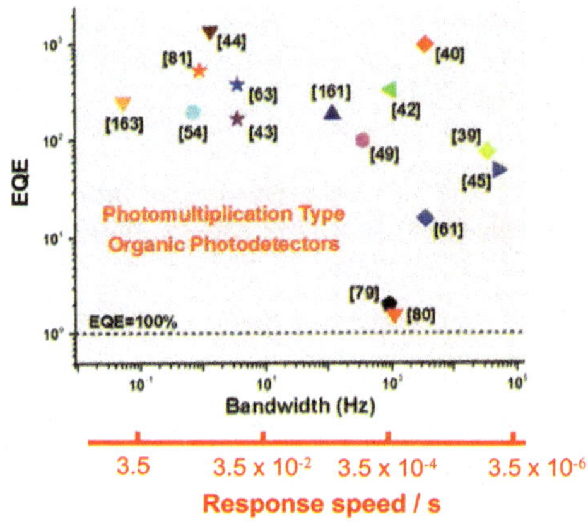

Fig. 11.9 Relationship between EQE and response speed (bandwidth) for OPM devices reported so far. *Note* The plot is logarithmic. EQE of 100% (10^0) is shown by broken line. Response speed (red horizontal axis) is approximately reciprocal to the bandwidth (black horizontal axis). Reproduced with the permission of [1]. © 2019, WILEY–VCH Verlag GmbH

11.10 Conclusion

The progress in photocurrent multiplication after 2000 was briefly summarized. The concept of organic photocurrent multiplication (OPM) was established. The first important development is the intentional formation of traps by uniform and interfacial doping, and incomplete percolation. The second important development is the intentional insertion of a blocking layer. The response speed of the multiplied photocurrent remains an issue in the construction of practical OPM devices.

References

1. Miao, J., Zhang, F.: Recent progress on photomultiplication type organic photodetectors. Laser Photonics Rev. **13**, 1800204 (38 pp) (2019)
2. Shi, L., Liang, Q., Wang, W., Zhang, Y., Li, G., Ji, T., Hao, Y., Cui, Y.: Research progress in organic photomultiplication photodetectors. Nanomaterials **8**, 713 (27 pp) (2018)
3. Hiramoto, M., Imahigashi, T., Yokoyama, M.: Photocurrent multiplication in organic pigment films. Appl. Phys. Lett. **64**, 187–189 (1994)
4. Hiramoto, M., Miki, A., Yoshida, M., Yokoyama, M.: Photocurrent multiplication in organic single crystals. Appl. Phys. Lett. **81**, 1500–1502 (2002)
5. Huang, J., Yang, Y.: Origin of photomultiplication in C_{60} based devices. Appl. Phys. Lett. **91**, 203505 (3 pp) (2007)
6. Dauber, T.K., Neher, D., Rost, H., Horhold, H.H.: Efficient bulk photogeneration of charge carriers and photoconductivity gain in arylamino-PPV polymer sandwich cells. Phys. Rev. B **59**, 1964–1972 (1999)
7. Kim, Y., Cook, S., Tuladhar, S.M., Choulis, S.A., Nelson, J., Durrant, J.R., Bradley, D.D.C., Giles, M., McCulloch, I., Ha, C.-S., Ree, M.: A strong regioregularity effect in self-organizing conjugated polymer films and high-efficiency polythiophene:fullerene solar cells. Nat. Mater. **5**, 197–203 (2006)

8. Osaka, I., Saito, M., Koganezawa, T., Takimiya, K.: Thiophene–thiazolothiazole copolymers: Significant impact of side chain composition on backbone orientation and solar cell performances. Adv. Mater. **26**, 331–338 (2014)
9. Li, G., Zhu, R., Yang, Y.: Polymer solar cells. Nat. Photonics **6**, 153–161 (2012) and the references therein
10. Hiramoto, M., Fujiwara, H., Yokoyama, M.: Three-layered organic solar cell with a photoactive interlayer of codeposited pigments. Appl. Phys. Lett. **58**, 1062–1064 (1992)
11. Hiramoto, M., Fujiwara, H., Yokoyama, M.: *p-i-n* like behavior in three-layered organic solar cells having c co-deposited interlayer of pigments. J. Appl. Phys. **72**, 3781–3787 (1992)
12. Matsunobu, G., Oishi, Y., Yokoyama, M., Hiramoto, M.: High-speed multiplication-type photodetecting device using organic co-deposited films. Appl. Phys. Lett. **81**, 1321–1322 (2002)
13. Yang, D., Zhou, X., Wang, Y., Vadim, A., Alshehri, A.M., Ahamadc, T., Ma, D.: Deep ultraviolet-to-NIR broad spectral response organic photodetectors with large gain. J. Mater. Chem. C **4**, 2160–2164 (2016)
14. Guo, D., Yang, D., Zhao, J., Vadim, A., Ma, D.: Role of interfaces in controlling charge accumulation and injection in the photodetection performance of photomultiplication-type organic photodetectors. J. Mater. Chem. C **8**, 9024–9031 (2020)
15. Dong, R., Bi, C., Guo, F., Yuan, Y., Fang, Y., Zhengguo, X., Huang, J.: An Ultraviolet-to-NIR Broad Spectral Nanocomposite Photodetector with Gain. Adv. Optical Mater. **2**, 549–554 (2014)
16. Melancon, J.M., Zivanovic, S.R.: Broadband gain on poly(3-hexylthiophene):phenyl-C61-butyric-acid-methyl-ester photodetectors enabled by a semicontinuous gold interlayer. Appl. Phys. Lett. **105**, 163301 (3 pp) (2014)
17. Miao, J., Zhang, F., Lin, Y., Wang, W., Cao, M., Li, L., Zhang, J., Zhan, X.: Highly sensitive organic photodetectors with tunable spectral response under bi-directional bias. Adv. Optical. Mater. **4**, 1711–1717 (2016)
18. Lin, Y.Z., Wang, J.Y., Zhang, Z.G., Bai, H.T., Li, Y.F., Zhu, D.B., Zhan, X.W.: An electron acceptor challenging fullerenes for efficient polymer solar cells. Adv. Mater. **27**, 1170–1174 (2015)
19. Jang, M.S., Yoon, S., Sim, K.M., Cho, J., Chung, D.S.: Spatial confinement of the optical sensitizer to realize a thin film organic photodetector with high detectivity and thermal stability. J. Phys Chem. Lett. **9**, 8–12 (2018)
20. Ishii, A., Sakai, T., Takanashi, R., Ogata, S., Kondo, K., Iwasawa, D., Mizushima, S., Yoshihara, K., Hasegawa, M.: Sensitive photodetection with photomultiplication effect in an interfacial $Eu^{2+/3+}$ complex on a mesoporous TiO_2 Film. ACS Appl. Mater. Interfaces **10**, 5706–5713 (2018)
21. Ishii, A., Jena, A.K., Miyasaka, T.: Photomultiplying visible light detection by halide perovskite nanoparticles hybridized with an organo Eu complex. J. Phys Chem. Lett. **10**, 5935–5942 (2019)
22. Benduhn, J., Tvingstedt, K., Piersimoni, F., Ullbrich, S., Fan, Y., Tropiano, M., McGarry, K.A., Zeika, O., Riede, M.K., Douglas, C.J., Barlow, S., Marder, S.R., Neher, D., Spoltore, D., Vandewal, K.: Intrinsic non-radiative voltage losses in fullerene-based organic solar cells. Nat. Energy **2**, 17053 (2017)
23. Izawa, S., Shintaku, N., Kikuchi, M., Hiramoto, M.: Importance of interfacial crystallinity to reduce open-circuit voltage loss in organic solar cells. Appl. Phys Lett. **115**, 153301 (4 pp) (2019)
24. Kublitski, J., Fischer, A., Xing, S., Baisinger, L., Bittrich, E., Spoltore, D., Benduhn, J., Vandewal, K., Leo, K.: Enhancing sub-bandgap external quantum efficiency by photomultiplication for narrowband organic near-infrared photodetectors. Nat. Commun. **12**, 4259 (9 pp) (2021)
25. Shen, L., Zhang, Y., Bai, Y., Zheng, X., Wang, Q., Huang, J.: A filterless, visible-blind, narrowband, and near-infrared photodetector with a gain. Nanoscale **8**, 12990–12997 (2016)